# The Sarcophagidae (Diptera)
# of Fennoscandia and Denmark

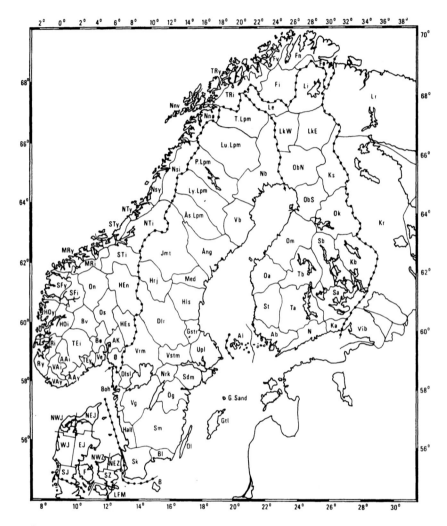

List of abbreviations for the provinces used throughout the text, on the map and in the following tables.

**DENMARK**

| | | | |
|---|---|---|---|
| SJ | South Jutland | LFM | Lolland, Falster, Møn |
| EJ | East Jutland | SZ | South Zealand |
| WJ | West Jutland | NWZ | North West Zealand |
| NWJ | North West Jutland | NEZ | North East Zealand |
| NEJ | North East Jutland | B | Bornholm |
| F | Funen | | |

# SWEDEN

| | | | |
|---|---|---|---|
| Sk. | Skåne | Vrm. | Värmland |
| Bl. | Blekinge | Dlr. | Dalarna |
| Hall. | Halland | Gstr. | Gästrikland |
| Sm. | Småland | Hls. | Hälsingland |
| Öl. | Öland | Med. | Medelpad |
| Gtl. | Gotland | Hrj. | Härjedalen |
| G. Sand. | Gotska Sandön | Jmt. | Jämtland |
| Ög. | Östergötland | Äng. | Ångermanland |
| Vg. | Västergötland | Vb. | Västerbotten |
| Boh. | Bohuslän | Nb. | Norrbotten |
| Dlsl. | Dalsland | Ås. Lpm. | Åsele Lappmark |
| Nrk. | Närke | Ly. Lpm. | Lycksele Lappmark |
| Sdm. | Södermanland | P. Lpm. | Pite Lappmark |
| Upl. | Uppland | Lu. Lpm. | Lule Lappmark |
| Vstm. | Västmanland | T. Lpm. | Torne Lappmark |

# NORWAY

| | | | |
|---|---|---|---|
| Ø | Østfold | HO | Hordaland |
| AK | Akershus | SF | Sogn og Fjordane |
| HE | Hedmark | MR | Møre og Romsdal |
| O | Opland | ST | Sør-Trøndelag |
| B | Buskerud | NT | Nord-Trøndelag |
| VE | Vestfold | Ns | southern Nordland |
| TE | Telemark | Nn | northern Nordland |
| AA | Aust-Agder | TR | Troms |
| VA | Vest-Agder | F | Finnmark |
| R | Rogaland | | |

n northern     s southern     ø eastern     v western     y outer     i inner

# FINLAND

| | | | |
|---|---|---|---|
| Al | Alandia | Kb | Karelia borealis |
| Ab | Regio aboensis | Om | Ostrobottnia media |
| N | Nylandia | Ok | Ostrobottnia kajanensis |
| Ka | Karelia australis | ObS | Ostrobottnia borealis, S part |
| St | Satakunta | ObN | Ostrobottnia borealis, N part |
| Ta | Tavastia australis | Ks | Kuusamo |
| Sa | Savonia australis | LkW | Lapponia kemensis, W part |
| Oa | Ostrobottnia australis | LkE | Lapponia kemensis, E part |
| Tb | Tavastia borealis | Li | Lapponia inarensis |
| Sb | Savonia borealis | Le | Lapponia enontekiensis |

# USSR

Vib    Regio Viburgensis    Kr    Karelia rossica    Lr    Lapponia rossica

FAUNA ENTOMOLOGICA SCANDINAVICA
Volume 19                                    1987

# The Sarcophagidae (Diptera) of Fennoscandia and Denmark

*by*

Thomas Pape

E. J. Brill / Scandinavian Science Press Ltd.

Leiden · Copenhagen

*Fauna entomologica scandinavica*
is edited by "Societas entomologica scandinavica"

*Editorial board*
Nils M. Andersen, Karl-Johan Hedqvist, Hans Kauri,
N. P. Kristensen, Harry Krogerus, Leif Lyneborg,
Hans Silfverberg

*Managing editor*
Leif Lyneborg

*World list abbreviation*
Fauna ent. scand.

*Colour separation*
Cromoscan, Rødovre, Denmark

*Printed by*
Vinderup Bogtrykkeri A/S
7830 Vinderup, Denmark


ISBN 90 04 08184.4
ISBN 87-87491-37-0
ISSN 0106-8377

Author's address:
Zoological Museum
Universitetsparken 15
DK-2100 Copenhagen
Denmark

# Contents

Plates 1 and 2 are arranged after p. 176.

# Introduction

The Sarcophagidae has for long been a neglected or superficially studied group amongst the Fennoscandian and Danish Diptera. Species recognition is difficult, and a definite identification can often only be made by a detailed study of the terminalia, frequently necessitating dissections; within a few species-groups, females are still impossible to identify.

A few Fennoscandian and Danish sarcophagids were described by Linnaeus (1758), De Geer (1776), and Fabricius (1787, 1794, 1805). Fallén (1810, 1817, 1820) and especially Zetterstedt (1838, 1844, 1845, 1859) added several more, but species-limits were still rather blurred. Starting with the work of Pandellé (1896), the male terminalia were recognised as providing very important diagnostic characters, and species-limits gradually stabilized. Lundbeck (1927) made a good attempt of recording, describing, and keying all the Danish species, but his species-concepts were entirely based on the existing and somewhat inadequate descriptions and illustrations by contemporary authors, and he missed several species. The classic work of Rohdendorf (1937), with excellent drawings of the male terminalia of almost every Palaearctic species of the Sarcophaginae then known, formed the basis of Tiensuu's (1939) revision of Finnish Sarcophagidae. The Finnish records have recently been updated in the checklists by Hackman (1980) and Pape (in press).

The Swedish Sarcophagidae, apart from the contributions of Linnaeus, De Geer, Fallén, and Zetterstedt mentioned above, have mainly been dealt with in a number of papers by Ringdahl, which culminated in his Catalogus Insectorum Sueciae XI (1952). In addition to a few misidentifications, Ringdahl failed to distinguish between the species of *Sarcophaga* sensu stricto.

Norwegian records of Sarcophagidae are rather few and scattered, but recently Rognes (1986a) has made a substantial contribution which greatly adds to our knowledge of the Fennoscandian Sarcophagidae.

All the papers mentioned above have almost entirely neglected the female sex within the Sarcophagini. The identification of female Sarcophagini does in fact involve great difficulties, and the present attempt to construct a usable key is far from successful, although *most* specimens of the *majority* of species may be identified.

The present work lists a total of 87 species from Fennoscandia and Denmark, and an attempt has been made to synthesize all the Fennoscandian and Danish records of the biology, e.g. behaviour, food-choice, hosts, etc. The catalogue showing the distribution is based mainly on specimens seen by the author, but records made by competent colleagues acquainted with the recent additions to the sarcophagid fauna have been included. All older records have been ignored as possible misidentifications. A few provincial records in Ringdahl (1952) are left unconfirmed, and at least one of these may be due to a slip on Ringdahl's part when preparing his catalogue: the record of *Sarcophaga hirticrus* from Öland is contradicted by Ringdahl (1937), who actually records the species from Gotland (supported by all the specimens in his collection).

The present state of knowledge of the Sarcophagidae from Fennoscandia and Denmark is still far from complete. A few more species will probably be found, and the presence of one species needs to be confirmed (= *Paramacronychia flavipalpis*). Much remains to be learned about distribution and biology.

# Acknowledgements

I wish to extend my sincerest thanks to Mr. K. Rognes, Madla, Norway, for his never-failing readiness to discuss the problems encountered during this work and for the prompt loan of material requested. I am indebted to Mr. R. Danielsson, Lund, for the loan of Sarcophagidae from the large Ringdahl collection, and to others who have provided me with material indispensable for the completion of this work: Mr. C. Bergström, Uppsala; Mr. P. Gjelstrup, Århus; Dr. B. Lindeberg, Helsinki; Mr. P. I. Persson, Stockholm; Mr. A. C. Pont, London; Mr. J. E. Raastad, Oslo; and Dr. Yu. G. Verves, Kiev. Dr. Yu. G. Verves kindly checked the identity of a few specimens and supplied me with important distribution data, and Mr. R. Richet, Boulogne, provided me with important new information on the nomenclature within *Sarcophaga* sensu stricto.

I wish to express my thanks to Mr. K. Pape for the skilled preparation of the colour plates, to all staff members of the entomological department of the Zoological Museum, Copenhagen, for advice, guidance, and good companionship during my work, and to the managing editor, Mr. L. Lyneborg, Copenhagen, for much support. Mr. A. C. Pont kindly checked the English of the manuscript.

The present work was sponsored by the Danish Natural Science Research Council.

Most of specimens studied for this work are deposited in the following institutions and private collections:
BMNH: British Museum (Natural History), London.
NMA: Naturhistorisk Museum, Århus.
NRM: Naturhistoriska Riksmuseet, Stockholm.
ZMH: Zoologiska Museet, Helsinki.
ZML: Zoologiska Museet, Zoologiska Institutionen, Lund.
ZMUC: Zoologisk Museum, Københavns Universitet, København.
Coll. Rognes: Private collection of K. Rognes, Madla.
Coll. Bergström: Private collection of C. Bergström, Uppsala.

# Diagnosis

The family Sarcophagidae contains about 2,500 described species and is worldwide in distribution. The greatest number of species is found in warmer climates, and species diversity decreases markedly in cold temperate and subarctic regions. Only a few species seem to be exclusively restricted to temperate regions.

The Sarcophagidae are typical members of the superfamily Tachinoidea (Tachinidae sensu lato of Griffiths (1972)), with a row of meral bristles, a forwardly-bent vein M, and an opercular metathoracic spiracle. The monophyly of the Sarcophagidae seems to be well-founded by at least three synapomorphies:

1) Uterus with a two-lobed incubatory pouch;
2) Alpha-setae absent from abdominal sternites;
3) Aedeagus with the dorsolateral processes fused.

The two-lobed uterus is parallelled in the larviparous Oestrinae (Catts 1964: fig. 3), but this may be convergence as species of the Hypodermatinae and the Gasterophilinae possess a simple uterus.

Alpha-setae, which seem to be of general occurrence in the Muscomorpha (McAlpine 1981), have not been found in any Sarcophagidae, but within the Tachinoidea most species of Calliphoridae and Rhinophoridae, and at least some species of Tachinidae (Lehrer 1973a) possess alpha-setae on one or more abdominal sternites.

The miltogrammatine type of aedeagus with fused dorsolateral processes, although often with a slightly desclerotised median stripe, may be a ground-plan feature of the Sarcophagidae. The aedeagal structure in the Paramacronychiinae and Sarcophaginae is highly apomorphic, but it seems to have developed from the condition found in the Miltogrammatinae.

These apomorphies are of little use for identification purposes when a large material is at hand, and the following combination of characters will separate any Danish or Fennoscandian Sarcophagidae from the remaining Tachinoidea:

Mouthparts well developed. Proanepisternum (propleuron) bare. Stem-vein bare. Lower calypters with inner margin following scutellum for a considerable distance before turning outwards rather abruptly. Subscutellum not greatly swollen, rarely slightly convex (*Nyctia*). Thorax never with long wavy golden hairs. Abdomen with or without silvery-grey pollinosity forming a more or less tessellate pattern, but never shining metallic blue or green. Abdominal sternite 2 never covered by lateral margins of tergite 1 + 2.

# Classification

The classification of Townsend (1937, 1938), which divided the Sarcophagidae into six tribes, was rather confusing and contributed little to the phylogeny of the family. The much more careful investigations of Rohdendorf (1937, 1967) led to the recognition of three major groups: Sarcophaginae, Paramacronychiinae (as Agriinae), and Miltogrammatinae, and a few genera of uncertain position: *Macronychia, Chrysogramma,* and *Sarcotachina,* which were also given subfamily rank. The classification of Kurahashi (1972, 1975) is identical with that of Rohdendorf, but Downes (1955, 1965) has put forward a slightly different view, with the Miltogrammatini (including *Macronychia*) and Paramacronychiini (as Agriini) treated as tribes of the subfamily Miltogrammatinae.

In the present work, the Sarcophagidae are divided into three subfamilies: Miltogrammatinae, Paramacronychiinae, and Sarcophaginae, and a sister-group rela-

11

tionship between the two latter is suggested. *Macronychia* (only genus of the Macronychiinae) has been included in the Miltogrammatinae, as a sister-group relationship between *Macronychia* and the remaining Miltogrammatinae seems to be unwarranted. The best argument for a sister-group relationship is the presence of relatively small (or "normal-sized") eyes in *Macronychia* compared with the enlarged eyes of other miltogrammatine species, but eye-size is obviously very variable and highly adaptive, according to mating and hunting strategies. Another character is the presence in *Macronychia* of two strong mid tibial ad bristles. Nearly all species of the Miltogrammatinae have only a single mid tibial ad bristle, while almost all Paramacronychiinae and Sarcophaginae have two. The single ad bristle is probably apomorphic within the Sarcophagidae, but a subdivision of the Miltogrammatinae based on this character would lead to paraphyletic groups.

Two apomorphies support the monophyly of Paramacronychiinae + Sarcophaginae: the fusion of T6 with syntergosternite 7 + 8, and the displacement of the acrophallus to the ventral surface of the aedeagus. Although Downes (1955) states that "evidence overwhelmingly indicates a very close affinity" between the Paramacronychiinae and Miltogrammatinae, he lists only one possible synapomorphy: the fusion of the gonopods with the hypandrium.

The genus *Eurychaeta* Brauer & Bergenstamm (*Helicobosca* Bezzi) has recently been transferred to the Calliphoridae (Rognes 1986b). I agree with this, and the genus is accordingly omitted from this volume.

No consistent classification based on the principles of phylogenetic systematics is available below the subfamily level, and the present trichotomous division of the Sarcophaginae into Raviniini, Protodexiini, and Sarcophagini is based mainly on Old World material.


# Morphology of adults

Adult Sarcophagidae are very similar to the Calliphoridae and Tachinidae in general appearance, and a detailed description of external characters is hardly required. All the general morphological terms used in this paper are illustrated in Figs 1-18, and the terminology follows that of McAlpine (1981), who has given an excellent account of the adult morphology of the Diptera. Most morphological terms are fully explained by these illustrations and need no further description, but some characters of great importance for identification purposes will be described below.

As the male and female terminalia provide the best (or only) characters for identification within several genera or species-groups, they will be discussed in more detail, especially the elaborate aedeagal structure of male Sarcophagini, which have given rise to an extensive terminology.

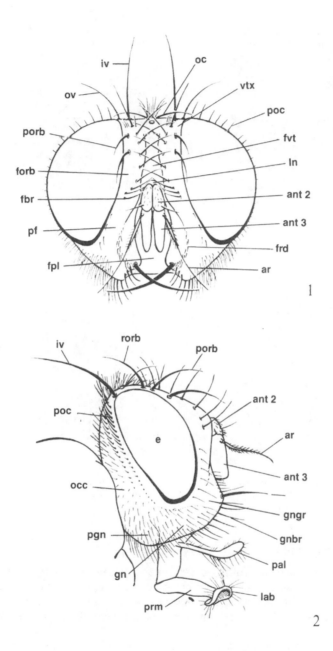

Figs 1, 2. Head. – 1: frontal view; 2: lateral view. Abbreviations on pp. 20, 21.

13

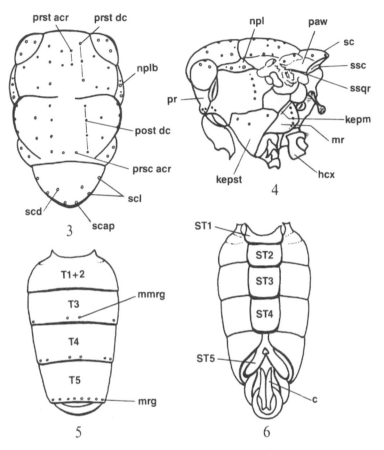

Figs 3-6. Thorax and abdomen. – 3: thorax, dorsal view; 4: thorax, lateral view; 5: abdomen, dorsal view; 6: abdomen, ventral view. Abbreviations on pp. 20, 21.

## Male terminalia

The Miltogrammatinae typically possess three-segmented male terminalia (Fig. 13), and only a few instances of reduction or partial fusion of the segments are known. The dorsal sclerites of the terminalia are the abdominal tergite 6, the composite syntergosternite 7 + 8, and the epandrium. Posterior to the epandrium are the paired cerci and surstyli.

Within the Paramacronychiinae and Sarcophaginae, a fusion of the first two segments has resulted in two-segmented terminalia, and most authors have used the terms

14

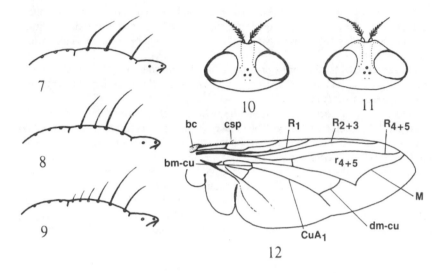

Figs 7-12. Details of adult morphology. – 7-9: different configurations of postsutural dorsocentral bristles; 10: head from above showing straight rows of frontal bristles; 11: head from above showing diverging rows of frontal bristles; 12: wing. Abbreviations on pp. 20, 21.

"first and second genital segment" in descriptive papers. In the present paper, the somewhat more informative term "protandrial segment" is used for the fusion product of T6 and syntergosternite 7 + 8, and the well-known term epandrium has been retained. The protandrial segment of the Paramacronychiinae is distinctly angulate in profile, and retains the row (rarely a tuft) of marginal bristles on T6 (Fig. 14).

The gonopods and parameres are rather uniform among closely related species, but may sometimes provide important diagnostic characters. It should be noted that the parameres are two-segmented, with a small basal sclerite. However, this is unimportant for identification purposes and the basal sclerite is omitted from the present figures.

The aedeagal structure of the Miltogrammatinae has probably not diverged markedly from the sarcophagid ground-plan (Fig. 16). The dorsolateral processes of the dorsal plate are fused on the dorsal median line, although a narrow median stripe of weaker sclerotisation often betrays its paired origin. The acrophallus, carrying the phallotreme, is terminal and is not distinctly set off from the rest of the aedeagus.

Within the Paramacronychiinae and Sarcophaginae, the acrophallus is much more well-defined and it has attained a position on the ventral surface of the aedeagus. Many amazing elaborations of aedeagal structure have arisen, especially within the Sarcophaginae, but only the acrophallus, with the phallotreme as a morphological landmark, has been successfully homologised throughout the family.

15

When dealing with the sparse Fennoscandian and Danish fauna, it is only the genera and species of the tribe Sarcophagini that require any detailed aedeagal descriptions, and the morphological terms applied within this group will be given particular attention.

The aedeagus of the Sarcophagini is articulated between basi- and distiphallus, and the latter possesses a number of sclerotisations of which the following are mentioned in the present paper (Figs 17, 18).

Juxta: the apical part of the distiphallus, which is often distinctly set-off from the basal part by a desclerotised strip. The juxta may have various paired structures, which have been given names of convenience: juxtal arms denote extensions of the basal margin found in *Parasarcophaga* and *Discachaeta,* juxtal processes arise from the apico-dorsal surface of the juxta in *Thyrsocnema,* and juxtal appendages are lobe-like extensions of the apico-lateral juxtal margin in some species of *Heteronychia.*

Styli: the acrophallus is completely divided into two lateral, more or less grooved sclerites, the styli, which guide the sperm into the ducts of the female seminal receptacles. The styli are variable in shape, but may be distinguished from the harpes by a

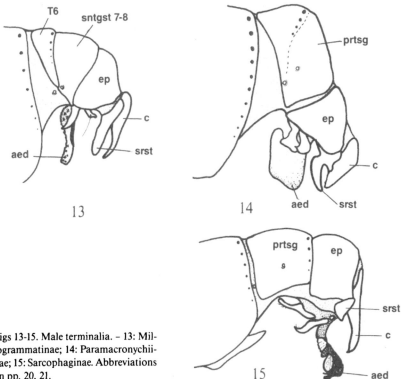

Figs 13-15. Male terminalia. – 13: Miltogrammatinae; 14: Paramacronychiinae; 15: Sarcophaginae. Abbreviations on pp. 20, 21.

16

microserration in the apical part (homologous to the recurved teeth of the miltogrammatine acrophallus).

Harpes: paired sclerotisations situated below the styli, often at the upper margin of the ventral plates. The harpes may be stylet-like or broadly flattened.

Median processes: many species of Sarcophagini possess a median sclerotisation on the ventral surface of the distiphallus, and this may be extended proximally into two curved median processes. The median processes are short but distinct in *Bellieriomima* and *Parasarcophaga,* and very long in *Thyrsocnema* (Figs 343, 348).

Vesica: a median structure situated between the ventral plates. The vesica may be unpaired or divided into 2-4 lobes.

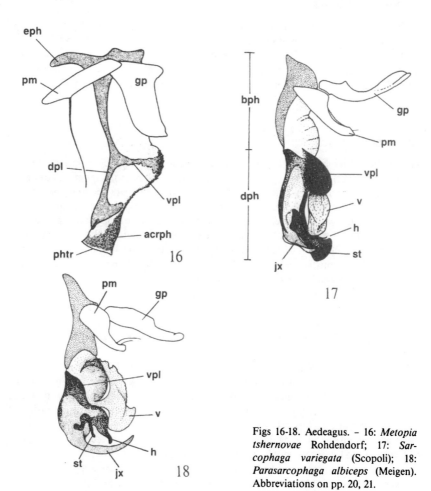

Figs 16-18. Aedeagus. – 16: *Metopia tshernovae* Rohdendorf; 17: *Sarcophaga variegata* (Scopoli); 18: *Parasarcophaga albiceps* (Meigen). Abbreviations on pp. 20, 21.

17

Ventral plates: ventral extensions of the dorsal sclerotisation. This structure is very distinct in the Miltogrammatinae, but may be difficult to delimit in many Sarcophaginae.

Many authors have coined their own terminology, or have modified that of earlier authors. Lehrer (1973b) has given a survey of the terminology of various authors, and additional references are Roback (1954), Lopes (1959), Kano *et al.* (1967), and Park (1977).

## Female terminalia

As in most other larviparous or ovo-larviparous flies, sarcophagid females possess shortened terminalia. The basic pattern consists of undivided tergites 6-8 with corresponding sternites, a pair of cerci, an epiproct, and a hypoproct, but there are various trends towards division, reduction, or fusion of the sclerites. An ovo-larvipositor is developed in some species of *Macronychia* (Miltogrammatinae) and in all species of *Blaesoxipha* (Sarcophaginae).

## Other characters

A character which has for the most part been overlooked is the distribution of black and white hairs on gena, postgena, and occiput within the Sarcophaginae. The demarcation between gena and postgena can be seen as a slightly depressed line (Fig. 2), and in many species of Sarcophaginae all the postgenal hairs, and only these, are white. In some species, the white hairs extend on to the posterior half of the gena (e.g. *Pierretia nigriventris* and *P. soror*), while others may have the gena almost wholly covered with white hairs (e.g. *Parasarcophaga albiceps* and *P. argyrostoma*). In *Pierretia sexpunctata,* almost all the postgenal hairs are black, and the white hairs are confined to the prestomal bridge below the neck.

Two other characters deserve mention. The chaetotaxy of the hind trochanters has hitherto been neglected in descriptive papers on the Sarcophagini. In the majority of species, both sexes possess a long seta apically on the median surface of hind trochanter, and males often possess specialised setae on the ventromedian part, e.g. short, dense bristles.

Many female Sarcophagini possess a mid femoral organ, visible as an oval elongate patch with reduced pollen and setulae on the posterior surface of the mid femur (Assis-Fonseca 1953). A similar structure may be present on fore femur, but then less developed. The mid femoral organ is situated at the middle or in the distal half, and the cuticle is more or less cross-striated by a number of widely-spaced grooves. Its function may be secretory, and Downes (1955) has reported that freshly-killed specimens may have numerous small droplets in this area.

The mid femoral organ may be bright red and conspicuous in some species, blackish and very difficult to detect in others, or totally absent, and a careful examination is necessary. The size and position of the mid femoral organ is reasonably constant within species, and even seems to be of use in the definition of larger groups.

# Morphology of immature stages

The larvae of most Sarcophagidae are typical maggots, with a tapering anterior end and a rounded or obliquely truncate posterior end. The morphological diversity is rather narrow, and the larvae of obligate parasites/parasitoids differ only slightly from the presumably plesiomorphic scavengers. The specialised pitcher-plant dwellers often have the rim encircling the posterior pair of spiracles ballooning outwards to form a float which suspends the larva from the surface film.

The larviparous (or ovo-larviparous) habit of all Sarcophagidae means that first instar larvae are readily accessible by abdominal dissection of pinned, dried females, and this stage is relatively well described. However, females are often difficult to identify, e.g. *Metopia* sensu stricto spp., *Agria* spp., and *Sarcophaga* spp., and knowledge is still much too limited to allow the construction of a key to species or even species-groups/genera.

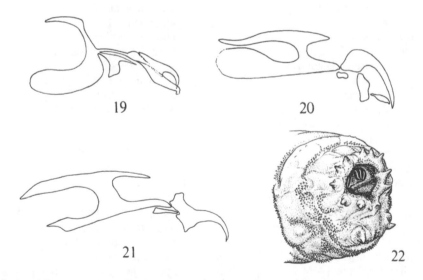

Figs 19-21. Cephalopharyngeal skeleton of first instar larva. – 19: *Miltogramma germari* Meigen; 20: *Nyctia halterata* (Panzer); 21: *Parasarcophaga argyrostoma* (Robineau-Desvoidy).
Fig. 22. Posterior end of third instar larva of *Ravinia* sp.

The sarcophagid larva, especially the third instar, is usually characterised by the posterior spiracular plate which has sunk into a deep pit which can be closed by pulling the upper and lower margin together (Fig. 22). Some species, however, have the posterior spiracular plate set normally on the body surface. The three spiracular slits of the third instar larva are more or less vertical, the peritreme surrounding the slits is usually incomplete, and the ecdysial scar from the spiracles of the preceding stage is indistinct.

The cephalopharyngeal skeleton of first stage larvae can be roughly divided into two types: the probably plesiomorphic type of the Miltogrammatinae and Paramacronychiinae, with a well-developed median sclerite and more or less well-developed paired mandibles (Figs 19, 20); and the type found in all Sarcophaginae, with a weakly sclerotised or reduced median sclerite and well developed mandibles (Fig. 21).

The puparium is of the typical barrel-shape found in many cyclorrhaphous flies. Apart from some shrinking of the posterior spiracular pit, which may render the spiracular slits difficult to see, most characters of the third instar larva remain distinct.

# Abbreviations for morphological terms

a = anterior
acrph = acrophallus
ad = anterodorsal
aed = aedeagus
ant2 = second antennomere
ant3 = third antennomere
ar = arista
av = anteroventral
bc = basicosta
bm-cu = basal medial-cubital crossvein
bph = basiphallus
c = cerci
csp = costal spine
$CuA_1$ = anterior branch of cubitus
dm-cu = discal medial-cubital crossvein
dph = distiphallus
dpl = dorsal plate
e = eye
ep = epandrium
eph = epiphallus
fbr = frontal bristles
forb = fronto-orbital plate
fpl = facial plate

frd = facial ridge
fvt = frontal vitta
gn = gena
gnbr = genal bristles
gngr = genal groove
gp = gonopod
h = harpes
hcx = hind coxa
iv = inner vertical bristle
jx = juxta
kepm = katepimeron
kepst = katepisternum
lab = labellum
ln = lunula
M = media
mmrg = median marginal bristles
mr = meron
mrg = marginal bristles
npl = notopleuron
nplb = notopleural bristles
oc = ocellar bristles
occ = occiput
ov = outer vertical bristle

p = posterior
pal = palpi
paw = post-alar wall
pd = posterodorsal
pf = parafacial plate
pgn = postgena
phtr = phallotreme
pm = paramere
poc = postocular setae
porb = proclinate orbital bristles
post dc = postsutural dorsocentral
  bristles
pr = proanepisternum
prm = prementum
prsc acr = prescutellar acrostichal
  bristles
prst acr = presutural acrostichal bristles
prst dc = presutural dorsocentral bristles
prtsg = protandrial segment

pv = posteroventral
$R_1$, $R_{2+3}$, $R_{4+5}$ = radial veins
$r_{4+5}$ = cell posterior to $R_{4+5}$
rorb = reclinate orbital bristles
sc = scutellum
scap = scutellar apical bristles
scd = scutellar discal bristles
scl = scutellar lateral bristles
sntgst 7-8 = syntergosternite 7+8
srst = surstyli
ssc = subscutellum
ssqr = suprasquamal ridge
st = styli
ST 1-5 = sternites 1-5
T 1-6 = tergites 1-6
v = vesica
vib = vibrissa
vpl = ventral plate
vtx = vertex

# Collecting, preserving and identification

## Collecting

Members of the subfamily Sarcophaginae are generally medium-sized to large, conspicuous flies, and they are active for most of the day (nocturnal or crepuscular species are not found in Fennoscandia or Denmark, although some species may accidentally come to light (Audcent 1951)). They are strong fliers; males will often gather on sheltered and sunny sites, and both sexes are attracted to the inflorescences of open flowers where they can be netted directly. Bait-traps using decomposing meat or faeces may capture both males and females of some species, while others may be bred from gastropods, earthworms, spider-eggs, or other insects.

The Miltogrammatinae are more rarely encountered in nature owing to their smaller size and less gregarious habits. They should be especially sought in sandy and heathy localities. Females may be netted while flying low above the ground in search for hosts or host nests, and males of a few species (e.g. *Metopia* spp.) may aggregate on leaves or stones. Malaise traps are very useful for catching miltogrammatine flies which kleptoparasitise terrestrial (ground-nesting) hosts.

## Preserving

Collected specimens should be pinned like other flies, but male specimens of the Sarcophaginae should have their terminalia extended, facilitating a definite identification. The extension can be done in several ways, e.g. with a curved needle, but it is often necessary to held the terminalia in the extended position while the specimen dries. It is usually easiest to extend the terminalia some hours after death, when the initial rigor has passed off, but it may be rather time-consuming or inconvenient to prepare a large material in the field. Dried specimens may subsequently be relaxed in a container with damp sand or blotting paper, and the addition of some crystalline menthol will prevent mould. Specimens a few years old are easy to handle after 1-2 days softening, and even very old specimens (50 years or more) may be treated with success. On the other hand, specimens caught in malaise traps using a surplus of ethyl acetate or similar solvent as killing agent may be brittle and almost impossible to soften, and these may be fully dissected as described below for the Miltogrammatinae.

Within the Miltogrammatinae, the terminalia may be extended as in the Sarcophaginae, but this is often not necessary for definite identification, and their small size and the much less sclerotised structure of the aedeagus makes this undesirable in most cases. If, for any reason, an investigation of the terminalia is required, the tip of abdomen may simply be cut off with a pair of small scissors and treated with hot 10% potassium hydroxide for about 5 minutes. After washing in alcohol, the terminalia may be studied in glycerol. While dissected terminalia of the Sarcophagidae may be glued to a piece of cardboard after washing in alcohol, miltogrammatine terminalia should be stored in glycerol in a microvial pinned with the specimen.

If cutting the abdomen is not desired, the entire abdomen may be broken off and treated in potassium hydroxide as described above. After dissecting out the terminalia, the abdominal pollinosity may be partly restored by careful washing in ethyl acetate.


## Identification

Most Miltogrammatinae and Paramacronychiinae are easily identified using the keys, and examination of the male terminalia will only be necessary in a few instances or in greased and incomplete specimens. Within the Sarcophaginae, the Raviniini is monotypic and will give no problems. The Protodexiini contains six species from Fennoscandia and Denmark and, when the terminalia are compared, only the males of *B. plumicornis* and *B. laticornis* may still give problems.

Members of the Sarcophagini are generally very difficult to identify without references to the terminalia. Males will give few problems when their terminalia are extended or dissected, but it should be noted that the dried aedeagus may be somewhat shrivelled, and if a detailed examination is needed then a glycerol preparation should be made. All drawings of the aedeagi in the present paper have been made from glycerol preparations unless otherwise stated.

As the copulatory apparatus is highly specific at the species level, identification of males may be most easily achieved by direct comparison with the figures, and then, if necessary, by consulting the key at an appropriate entry.

When identifying females, many pitfalls exist which may lead to a wrong result, and females caught in copula should always be pinned with their males as they are valuable reference specimens. The mid femoral organ may easily be overlooked by the inexperienced observer, and the presence or absence of thoracic and abdominal bristles is a somewhat unreliable character in many species.

# Biology

Information on the biology of flesh-flies is still very sparse and, for most species, restricted to isolated breeding records. The biology of many species is still completely unknown. One feature of the sarcophagid ground-plan seems to have had important biological consequences: the in-utero incubation of the eggs. All sarcophagids either deposit incubated eggs with mature first instar larvae ready to hatch, or newly-hatched first instar larvae, or second instar larvae which have been nourished by the maternal accessory glands. The greater maternal investment in the progeny has evolved at the expense of a high fecundity. The Sarcophagidae probably arose from an ancestor which bred in dead animal matter, and several species in the recent fauna are necrophagous. Compared with the Calliphoridae, however, there has been a specialisation towards small bodies of carrion (Beaver 1977; Hanski & Kuusela 1980), probably because the habit of larviposition gives the larvae an advantage in exploiting a small food source and a more temporary environment. Whether or not larviposition evolved as a "small-carrion strategy", it seems to have been a pre-adaptation for the successful exploitation of living gastropods, earthworms, and insects.

In the case of snails, slugs, and earthworms, the fly larvae are scavengers or predators, and they kill and eat their prey before pupation. With regard to insect prey, all transitions from predators to parasitoids are found, and some species are even close to true parasites, as the prey/host may survive and feed after the escape of the larva. In general prey/host specificity seems to be low or moderate, e.g. species of *Sarcophaga* are predators of earthworms of the family Lumbricidae and *Agria punctata* parasitises a broad array of butterfly and moth genera. Some species, however, may be highly host-specific, e.g. *Agria mamillata* which preys upon larvae and pupae of ermine moths (*Yponomeuta* spp.), and *Blaesoxipha auditrix,* a Canadian parasitoid of male cicadas, which locates the host by its song (Soper *et al.* 1976).

Within the genus *Blaesoxipha* sensu lato, the shape of the female larvipositor may reflect the degree of host-specificity.

Some species have specialised in the egg-sacs of spiders, and others deposit their larvae in the pitchers of pitcher-plants (*Sarracenia* and *Nepenthes*), where they feed on the insects trapped in the proteolytic fluid (Forsyth & Robertson 1975; Beaver 1979).

A large group of sarcophagids, the subfamily Miltogrammatinae, are kleptoparasites of solitary wasps and bees. The female flies deposit their larvae or incubated eggs on the prey or pollen-balls provisioned for the "host" progeny, or they follow the host returning with its prey and glue some eggs directly on to the prey or the body of the host. This act may be performed on hosts dragging the prey to the nest as well as on hosts carrying the prey in flight. In most instances, the fly larva will devour the egg or young larva of the hymenopteran or the latter will die from starvation. In host-species which continually provision the nest both fly and hymenopteran may develop to maturity.

Species of Miltogrammatinae generally possess very large eyes and females may have the front facets distinctly enlarged, facilitating the detection and pursuit of potential hosts. When a female fly detects a passing object which fits the behaviour and size-range of the host-species, it will follow close behind it. The name "satellite-flies" (mostly applied to species of *Senotainia*) is due to this habit of following the host at a rather well-defined distance. If the passing object turns out to be a suitable host, the fly may larviposit in flight or follow the host to the nest. In the latter instance, larviposition may take place when the host drags the prey into the burrow, or the fly may wait patiently for the return of the host, after which it quickly slips into the burrow, deposits one or more larvae, and escapes before the host has closed the entrance.

In some cases the female fly may larviposit directly into the entrance, and the larvae will actively seek the prey (Ristich 1956; Endo 1980). Some species of Miltogrammatinae search the ground for the open entrances of recently-provisioned nests, but others may re-excavate closed entrances with their flattened fore tarsi (Allen 1926). The latter applies to species of *Phrosinella*, but detailed investigations are needed. Females in other genera have their fore tarsi similarly flattened, but their biology is still unknown.

The complex behavioural diversification regarding transport of prey, closing of nest entrances, and male investment within solitary wasps and bees may be partly due to kleptoparasitising Miltogrammatinae (Evans 1970; Evans & Eberhard 1970; Peckham 1977).

A few species of Miltogrammatinae have been recorded as insect parasitoids, and Verves (1976a) considers kleptoparasitism to be a secondary development from insect parasitism. It should be noted, however, that the larvae seem to be more or less external to their prey, acting more like scavengers than parasitoids, and it is more probable that kleptoparasitism evolved as a "small-carrion strategy", with the habit of ovo-larviposition as an important pre-adaptation.

The larvae of a few sarcophagid species, mostly from the subfamily Miltogrammatinae, live in termite mounds. One of these, *Termitometopia skaifei*, is fed with regurgitated food by the termite workers in return for some substance produced (Skaife 1955).

Some species of Sarcophagidae attack vertebrates, and several instances of secondary myiasis in sheep and the infestation of necrotic wounds are known, although the Sarcophagidae are much less important members of the myiasis-producing community than the Calliphoridae. Only a few species seem to be obligate parasites of vertebrates: some attack reptiles (Dodge 1955), but the most notorious vertebrate parasites are species of *Wohlfahrtia* which may cause serious damage to their host (Zumpt 1965).

# Key to genera of Sarcophagidae

1　Hind coxae bare on posterior surface. Notopleuron with 2 strong bristles and with or without additional hairs. Abdominal ST3-ST4 more or less concealed by tergal margins in both sexes .............................................. 2

－　Hind coxae with fine hairs on posterior surface (Fig. 4). Notopleuron with 2 strong primary bristles, 2 smaller subprimary bristles, and with or without additional hairs (Figs 3,4). Males with abdominal ST3-ST4 fully exposed and overlapping tergal margins (Fig.6) ........................... 19

2(1)　Mid tibia with at least 2 strong ad bristles......................... 3

－　Mid tibia with only 1 ad bristle, or without any ad bristles ............ 10

3(2)　Vibrissae situated above lower margin of facial plate. Vibrissal angle not projecting (Fig. 162)..... *Macronychia* Rondani (p. 78)

－　Vibrissae, if present, situated at about lower margin of facial plate. Vibrissal angle slightly projecting ....................... 4

4(3)　Vibrissae not differentiated. Eyes large (Fig. 23). 3-5 proclinate orbital bristles................. *Miltogramma* Meigen (p. 27)

－　Vibrissae well developed. Eyes not enlarged. 0-2 proclinate orbital bristles.................................................. 5

5(4)　Arista bare or short-haired, hairs distinctly shorter than greatest aristal diameter......................................... 6

－　Arista plumose, hairs distinctly longer than greatest arital diameter ..................................................... 7

6(5)　Third antennomere 1.0-1.4× as long as second. Gena at least 0.5×eye-height. ♂: aedeagus as in Fig. 205. ♀: abdominal T6 well developed, distinctly visible, and densely pollinose.......... *Paramacronychia* Brauer & Bergenstamm (p. 96)

－　Third antennomere 1.7-2.2× as long as second. Gena at most 0.5×eye-height. ♂: aedeagus as in Fig. 199. ♀: abdominal T6 more or less concealed below T5, divided along the dorsal median line, and weakly pollinose....................
...................................... *Brachicoma* Rondani (p. 92)

7(5)　Wings infuscated along anterior margin. Basicosta mostly black. Abdomen black and almost unpollinose .......................
.................................. *Nyctia* Robineau-Desvoidy (p. 95)

－　Wings hyaline. Basicosta yellow. Abdomen with dense pollinosity ................................................. 8

8(7)　Costal spine not differentiated ........ *Agria* Robineau-Desvoidy (p. 84)

－　Costal spine well developed ....................................... 9

9(8)　Palpi black. Frons broad, 0.40-0.50×head-width, and with proclinate orbital bristles in both sexes ... *Sarcophila* Rondani (p. 89)

－　Palpi yellow. Frons narrower, in ♂ 0.19-0.23×head-

25

width, in ♀ 0.33-0.40 × head-width. Males without proclinate orbital bristles ......... *Angiometopa* Brauer & Bergenstamm (p. 90)
10(2) Numerous hair-like proclinate orbital bristles. Head profile rounded (Fig. 77) ............... *Amobia* Robineau-Desvoidy (p. 45)
– 2-5 strong proclinate orbital bristles. Head profile different .......... 11
11(10) Wing cell $r_{4+5}$ closed at wing-margin or with a very short
petiole ..................................................................... 12
– Wing cell $r_{4+5}$ open at wing-margin ........................... 13
12(11) Arista bare.............................. *Taxigramma* Perris (p. 61)
– Arista short-haired, longest hairs slightly longer than
greatest aristal diameter ..................... *Hilarella* Rondani (p. 57)
13(11) Lunula with setae .............................................. 14
– Lunula without setae ......................................... 15
14(13) Parafacial plate with a row of bristles along inner margin,
close to facial ridge (Figs 135-138). Palpi black .... *Metopia* Meigen (p. 69)
– Parafacial plate without bristles along inner margin. Pal-
pi yellow ..................... *Phrosinella* Robineau-Desvoidy (p. 66)
15(13) 3-5 strong proclinate orbital bristles ........................... 16
– 2 proclinate orbital bristles ...................................... 17
16(15) Arista more or less flattened (Figs 82, 83). Males with
spotted wings ............................. *Phylloteles* Loew (p. 49)
– Arista not flattened (Fig. 23). Wings hyaline in both sexes.
.................................................. *Miltogramma* Meigen (p. 27)
17(15) Vibrissal angle retreating, much less prominent than frons
(Figs 84-86) ....................... *Oebalia* Robineau-Desvoidy (p. 51)
– Vibrissal angle as prominent as frons or only slightly less
prominent (Figs 54, 59-61) ....................................... 18
18(17) Notopleuron with only 0-3 hairs in addition to the usual
2 bristles. Claws of fore legs in ♂ 1.0-1.2×, and in ♀ 0.75
×, as long as fifth tarsomere .............. *Senotainia* Macquart (p. 40)
– Notopleuron with 5-20 hairs in addition to the usual 2
bristles. Claws of fore legs in ♂ 0.75-0.80×, and in ♀
0.5-0.7×, as long as fifth tarsomere ... *Pterella* Robineau-Desvoidy (p. 38)
19(1) Row of frontal bristles more or less straight in dorsal view
(Fig. 10) ........................ *Ravinia* Robineau-Desvoidy (p. 99)
– Row of frontal bristles distinctly curving outwards at lunula (Fig. 11)................................................... 20
20(19) ♂: cerci distinctly bent dorsally (Figs 210, 232). ♀: abdominal ST7-ST8 fused and forming a sclerotised larvipositor which may be shovel-shaped, blade-like, or recurving beneath the abdomen (Figs 212, 230, 234) . *Blaesoxipha* Loew (p. 100)
– ♂: cerci straight, not bent dorsally. ♀: terminalia not
modified into a larvipositor; ST7-ST8 not fused. .....................
.............................. all genera of the Sarcophagini (p. 112)

## SUBFAMILY MILTOGRAMMATINAE

Generally small to medium-sized species, but a few species from the warmer parts of the Old World may be rather large (*Krombeinomyia, Hoplacephala* (biology unknown)).

Eyes large, except in most species of *Macronychia,* and the front facets sometimes greatly enlarged in females. Frons equibroad and with proclinate orbital bristles in both sexes. Arista bare to micropubescent, rarely short-haired or plumose. Notopleuron with 2 strong bristles, and with or without additional hairs. Katepisternum usually with 2 bristles, rarely 2:1 or 3:1. Katepimeron separated from meron (coxopleural streak present). Mid tibia usually with 1 ad bristle, seldom with 2. Hind coxae bare on posterior surface. Calypters enlarged. Male terminalia with 3 segments. Aedeagus with a single, undivided dorsal plate and terminal phallotreme. Epiphallus usually present.

The species are kleptoparasites in the nests of solitary wasps and bees, but insect parasitoids and termite inquilines have been recorded, and in North America the larva of *Eumacronychia sternalis* has been recorded as a predator of sea-turtle eggs and hatchlings (Lopes 1982).

## Genus *Miltogramma* Meigen, 1803

*Miltogramma* Meigen, 1803, Mag. Insektenk., 2: 280.
Type species: *Miltogramma punctatum* Meigen, 1824.

Small to medium-sized species with a rather large yellow head. Parafacial plate broad. Vibrissal angle well above lower facial margin. Vibrissae not differentiated, equal in size to other setae on facial ridge and genal setae. Frontal bristles numerous; orbital bristles varying in number, often 3-4 proclinate and 1 reclinate. Parafacial plate bare or with fine hyaline hairs. Proboscis of medium length or long. Fore tarsus often with a short first tarsomere. Mid tibia with more than one ad bristle. Wings somewhat shining due to reduced clothing-hairs; cell $r_{4+5}$ open.

The genus *Miltogramma,* as defined here, is large and widely distributed in the Old World. Some authors prefer a division into several genera (e.g. Rohdendorf 1970; Rohdendorf & Verves 1980; Verves 1986).

### Key to species of *Miltogramma*

1  Suprasquamal ridge setose. Abdominal pattern with 3 well-
   defined black spots on each of T3-T5. Fourth tarsomere
   of male fore tarsus with long and dense hairs on anterior
   surface and 6 very long p bristles (Fig. 26) . . . . . . . . . . . 5. *punctatum* Meigen
-  Suprasquamal ridge bare. Abdominal pattern without
   spots or with a pattern which changes with the incidence
   of light . . . . . . . . . . . . . . . . . . . . . . . . . . . . . . . . . . . . . . . . . . 2

2(1)  Abdominal tergites black. T3-T5 each with silvery polli-
       nosity in anterior 0.4-0.6 except for a median non-polli-
       nose stripe. Proboscis short, prementum about as long as
       palpi. Male fore tarsus unmodified ................ 3. *ibericum* Villeneuve
   -   Abdominal T3-T5 entirely covered with pollinosity.
       Pattern unicolorous grey or distinctly changing with the
       incidence of light ................................................... 3
3(2)  Occipital setae all white except for the post-ocular setae
       which are black. Male fore tarsus with a pair of long setae
       on apicodorsal surface of tarsomeres 1-3 (Fig. 28) ......................
       ......................................... 6. *testaceifrons* (von Roser)
   -   Occipital setae all black, but genal and postgenal hairs
       sometimes white ................................................. 4
4(3)  Proboscis short, prementum 1.0-1.2 × length of palpi. Male
       fore tarsus with 1-2 long hairs on apical part of pd and ant-

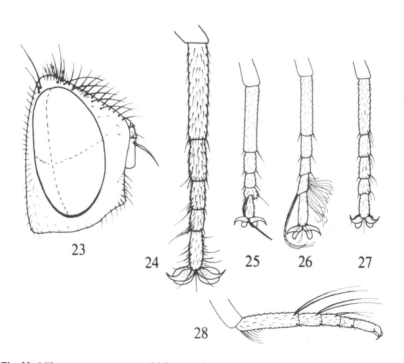

23

24    25    26    27

28

Fig. 23. *Miltogramma punctatum* Meigen, ♂ head.
Figs 24-27. Right fore tarsus ♂, dorsal view. – 24: *M. brevipilum* Villeneuve; 25: *M. oestraceum* (Fallén); 26: *M. punctatum* Meigen; 27: *M. villeneuvei* Verves.
Fig. 28. *Miltogramma testaceifrons* (von Roser), right fore tarsus ♂, posterior view.

erior surfaces of tarsomeres 1-4 (Fig. 27) . . . . . . . . . . . . . 7. *villeneuvei* Verves
- Proboscis long, prementum 1.5-2.0 × length of palpi. Male
  fore tarsus different or unmodified . . . . . . . . . . . . . . . . . . . . . . . . . . . . . . . 5
5(4) Fronto-orbital plate densely haired along entire length.
  Male fore tarsus with 2-3 elongate pd hairs on each of tar-
  someres 3-5 (Fig. 24) . . . . . . . . . . . . . . . . . . . . . . . . 1. *brevipilum* Villeneuve
- Fronto-orbital plate haired at vertex, bare or with only a
  few hairs on lower part. Male fore tarsus different . . . . . . . . . . . . . . . . . . . 6
6(5) Abdomen unicolorous grey, the reflections only slightly
  changing with the incidence of light. Arista thickened in
  proximal 0.4-0.6. Fourth tarsomere of male fore tarsus with
  a long ad seta and a tuft of long flattened pv bristles which
  projects between the claws (Fig. 25). . . . . . . . . . . . . . . 4. *oestraceum* (Fallén)
- Abdomen yellowish brown to blackish brown, pattern
  distinctly changing with the incidence of light. Arista
  thickened in proximal 0.5-0.8. Male fore tarsus unmodified

. . . . . . . . . . . . . . . . . . . . . . . . . . . . . . . . . . . . . . . . . . . . . . . . . . . 2. *germari* Meigen

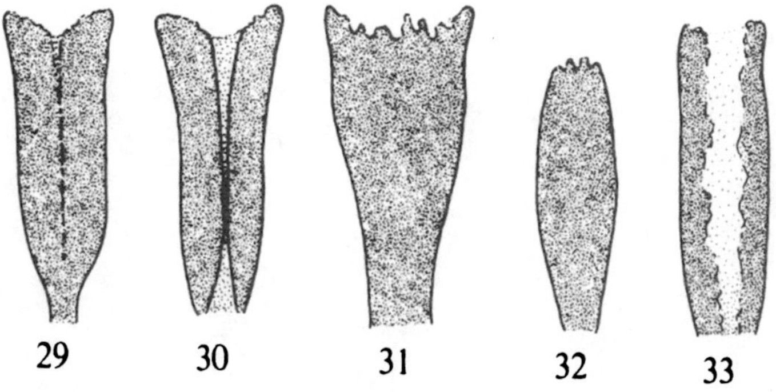

Figs 29-33. *Miltogramma* spp., dorsal plate of aedeagus. – 29: *M. brevipilum* Villeneuve; 30: *M. germari* Meigen; 31: *M. oestraceum* (Fallén); 32: *M. punctatum* Meigen; 33: *M. villeneuvei* Verves.

## 1. *Miltogramma brevipilum* Villeneuve, 1911
Figs 24, 29, 34-37.

*Miltogramma brevipilum* Villeneuve, 1911, Dt. ent. Z., 1911: 118.

Grey species. Fronto-orbital plates densely haired. Male fore tarsus with elongate pd hairs on tarsomeres 3-5. Abdomen ash-coloured, colour somewhat changing with the incidence of ligth.

♂. Head yellow. Frons almost parallel, frontal vitta slightly narrowing towards lunula. Fronto-orbital plate with 4 (3-5) proclinate and 1 reclinate orbital bristles and

29

densely haired along entire length. Parafacial plate sparsely haired on upper half or completely bare. First and second antennomeres light brown, third antennomere blackish grey and 1.9-2.2× as long as second. Arista thickened in proximal 0.6. Palpi yellow. Proboscis long. Thorax and legs brownish grey pollinose. Fore tarsus with 2-3 elongate pd hairs on each of tarsomeres 3-5. Fifth tarsomere in addition with some long hairs on anterior surface (Fig. 24). Abdomen with ash-coloured pollinosity with an indication of a median stripe. Colour pattern slightly changing with the incidence of light. T3 with an inconspicuous pair of median marginal bristles, T4-T5 with a row of marginals. Terminalia: surstyli with a characteristic hump at base (Fig. 34). Dorsal plate of aedeagus somewhat like *M. germari* (Fig. 29).

♀. Like the male but with unmodified fore tarsus.

Length ♂♀. 5.5-7.5 mm.

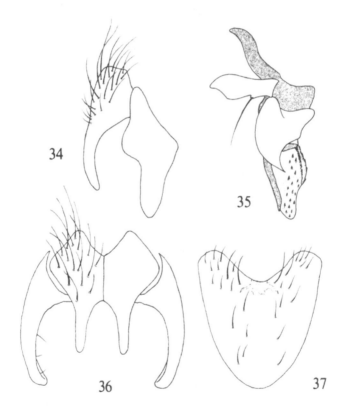

Figs 34-37. *Miltogramma brevipilum* Villeneuve. – 34: cerci + surstyli, lateral view; 35: aedeagus; 36: cerci + surstyli, posterior view; 37: ST5 ♂.

Distribution. Not recorded from Denmark and Finland. In Norway a single female from HE, and in Sweden known from Öl. and Gtl. – Western Europe east to Turkestan. Not in the British Isles.

Biology. Unknown.

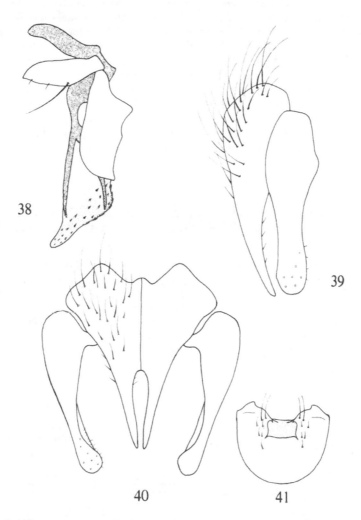

Figs 38-41. *Miltogramma germari* Meigen. – 38: aedeagus; 39: cerci + surstyli, lateral view; 40: cerci + surstyli, posterior view; 41: ST5 ♂.

## 2. *Miltogramma germari* Meigen, 1824
Figs 30, 38-41.

*Miltogramma germari* Meigen, 1824, Syst. Beschr., 4: 229.

A dark species with broad frons and greyish brown abdominal pattern which changes with the incidence of light. Male fore tarsus without specialised hairs.

♂. Head yellow. Fronto-orbital and parafacial plates light yellow pollinose, vertex darker and with sparser pollinosity. Ocellar triangle black. Frontal vitta almost parallel. Frontal bristles irregularly spaced on upper part, 4 (3-5) proclinate and 1 reclinate orbital bristles. Fronto-orbital plate with several rather long hairs at vertex, only a few hairs on lower part. Parafacial plate bare. First and second antennomeres and base of third orange, third antennomere 1.7-2.0× as long as second. Arista thickened in proximal 0.5-0.8. Palpi yellow. Proboscis long. Thorax and legs black with grey pollinosity. Fore tarsus without specialised hairs. Abdomen yellow brown to blackish brown, grey pollinose. Pattern of ill-defined blackish brown spots changing with the incidence of light. T1 + 2 & T3 brown laterally. A pair of median marginal bristles on T1 + 2 (often weak) and T3. T4-T5 with a row of marginals. Terminalia: aedeagus with long ventral plates and a longitudinally divided dorsal plate (Figs 30, 38).

♀. Like the male.
Length ♂♀. 6.0-8.5 mm.

Distribution. Common in Denmark and southern parts of Sweden and Finland. Not recorded from Norway. – Palaearctic: from the British Isles to Siberia and Mongolia. North Africa.

Biology. Recorded from nests of Apidae: *Anthophora, Megachile.*

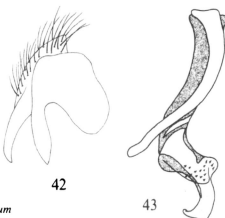

42

Figs 42, 43. *Miltogramma ibericum* Villeneuve. - 42: cerci + surstyli, lateral view; 43: aedeagus.

43

3. *Miltogramma ibericum* Villeneuve, 1912
   Figs 42, 43.

*Miltogramma ibericum* Villeneuve, 1912b, Bull. Mus. natn. Hist. nat. Paris, 1912: 508.
*Cylindrothecum necopinatum* Rohdendorf, 1930, 64h. Sarcophaginae, Fliegen pal.
   Reg., 11: 31.

A blackish species with proboscis of medium length. Male fore tarsus without specialised hairs. Abdominal tergites black with grey pollinose transverse bands interrupted
by a median black stripe.
   ♂. Head yellow. Frons and frontal vitta almost parallel. Fronto-orbital plate densely haired at vertex, more sparsely in anterior part, and with 3-7 orbital bristles of which
the upper 3-5 are stronger than the frontals. Parafacial plate with hyaline hairs in lower
parts. Apical part of second antennomere and base of third yellowish-brown, antennae otherwise brown or blackish brown. Third antennomere 1.4-1.8× as long as second. Arista thickened in proximal 0.4-0.6. Palpi yellow. Proboscis of medium length.
Thorax and legs black with grey pollinosity. Fore tarsus without specialised hairs. Abdomen black. T4-T5 each with a transverse band of grey pollinosity which is interrupted by a median black stripe. Terminalia: parameres reduced, gonopods elongated. Aedeagus highly characteristic, with a very long basiphallus (Fig. 43).
   ♀. Like the male.
   Length ♂♀. 7.5-9.0 mm.

Distribution. Recorded from Finland: Al, but not found elsewhere in Fennoscandia
or Denmark. – Widely distributed in the Palaearctic region. Probably in the Oriental
region as well, but most records seem to refer to the closely related *Miltogramma angu-
stifrons* (Townsend). Not in the British Isles.

Biology. Unknown.


4. *Miltogramma oestraceum* (Fallén, 1820)
   Figs 25, 31, 44-47; pl. 1:1.

*Tachina oestracea* Fallén, 1820, Monogr. Musc. Sveciae: 17.

Grey species. Frons broad at vertex and distinctly narrowing towards lunula. Fourth
tarsomere of male fore tarsus with a long ad seta and a tuft of long flattened pv bristles
which projects between the claws. Abdomen grey, slightly changing in colour with the
incidence of light.
   ♂. Head yellow. Frons narrowing towards lunula. Frontal vitta broad, almost parallel. Frontal bristles irregularly spaced on upper part. Fronto-orbital plate with 4 (3-5)
proclinate and 1 reclinate orbital bristles, with several long recurved hairs at vertex,
and more sparsely haired on lower part. Parafacial plate bare. First and second antennomeres orange, third antennomere grey to reddish grey and 1.8-2.4× as long as second. Arista thickened in proximal 0.4-0.6. Palpi yellow. Proboscis long. Thorax and
legs grey pollinose. Fore tarsus with an asymmetrical fourth tarsomere with a long ad

seta and a ventral tuft of five flattened and closely adpressed bristles which projects between the claws (Fig. 25). Abdomen densely grey pollinose without any trace of spots, but the colour changing slightly with the incidence of light. T1 + 2-T3 with a pair of median marginal bristles (often weak on T1 + 2). T4 with a row of marginals. Terminalia: aedeagus with the dorsal plate broadening apically (Fig. 31).

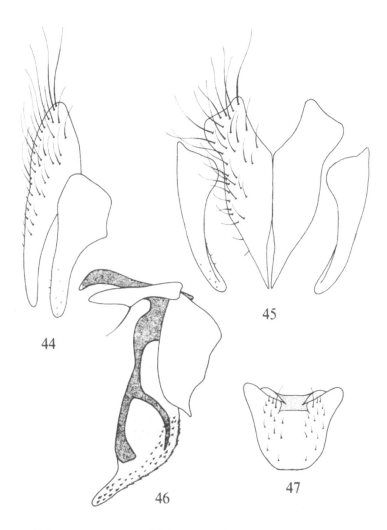

Figs 44-47. *Miltogramma oestraceum* (Fallén). – 44: cerci + surstyli, lateral view; 45: cerci + surstyli, posterior view; 46: aedeagus; 47: ST5 ♂.

34

♀. Very like the male but without specialised hairs on fore tarsus.
Length ♂♀. 7.0-8.5 mm.

Distribution. Occurring in Denmark, and in Sweden north to Nrk. Not recorded from Norway. In Finland only in the southern-most provinces. – Widely distributed in the Palaearctic region, from western Europe and North Africa east to Siberia and Central Asia. Not in the British Isles.

Biology. Recorded from nests of Sphecidae: *Cerceris, Podalirius;* Apidae: *Anthophora, Megachile, Dasypoda.*

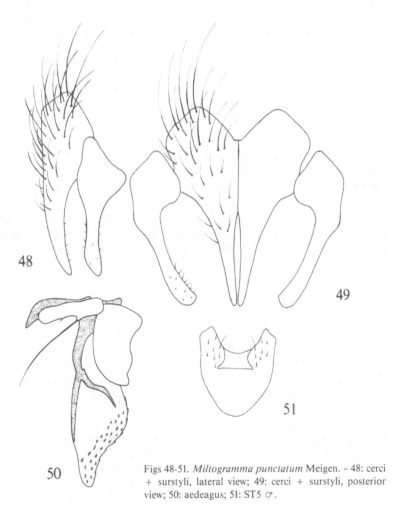

48

49

51

50

Figs 48-51. *Miltogramma punctatum* Meigen. – 48: cerci + surstyli, lateral view; 49: cerci + surstyli, posterior view; 50: aedeagus; 51: ST5 ♂.

## 5. *Miltogramma punctatum* Meigen, 1824
Figs 23, 26, 32, 48-51; pl. 1:2.

*Miltogramma punctata* Meigen, 1824, Syst. Beschr., 4: 228.

Easily distinguished from other Fennoscandian and Danish species by narrow frons, narrowing frontal vitta, haired supra-squamal ridges, and well-defined abdominal spots. Male with characteristic hairs on the fore tarsus.

♂. Head yellow. Fronto-orbital and parafacial plate light yellow pollinose, frontal vitta more golden-yellow pollinose. Fronto-orbital plate with 3-4 proclinate and 1 reclinate orbital bristles, with long reclinate hairs on vertex, and with fine hairs on anterior part. Parafacial plate bare. First and second antennomeres brown, the latter with a yellow apical margin. Third antennomere brownish grey, 1.5-2.0× as long as second. Arista thickened in proximal 0.5. Palpi yellow. Proboscis long. Thorax and legs black, grey pollinose. Supra-squamal ridge with a dense row of hairs. Fourth tarsomere of fore tarsus with long and dense hairs on anterior surface and 6 very long p bristles (Fig. 26). Abdomen brown with grey pollinosity. Tergites with 3 black spots. T3-T4 with a row of marginal bristles. Terminalia: aedeagus with an oval dorsal plate (Fig. 32).

♀. Like the male but with unmodified fore tarsus.

Length ♂♀. 5.5-9.5 mm.

Distribution. Common in Denmark. In Fennoscandia north to VA in Norway, Upl. in Sweden, and Om in Finland. – Widely distributed in the Palaearctic region, from western Europe to Siberia, Mongolia, and Japan; south to North Africa and the Canary Islands.

Biology. Recorded from nests of Sphecidae: *Ammophila, Trachelioides;* Apidae: *Colletes, Halictus.* Bred from *Colletes succincta* (Linnaeus) and *Halictus sexnotatulus* Nylander in Finland (Tiensuu 1939).

## 6. *Miltogramma testaceifrons* (von Roser, 1840)
Fig. 28.

*Xysta testaceifrons* von Roser, 1840, CorrespBl. württ. landw. Ver. Stuttg. (N. S.), 17(1): 57.
*Miltogramma pilitarse* Rondani, 1859, Dipt. ital. Prodromus, 3: 218.

Grey species. Most occipital setae white. Male fore tarsus with a pair of elongate hairs on tarsomeres 1-3. Abdomen grey, slightly changing in colour with the incidence of light, and with a blackish grey median stripe.

♂. Head yellow. Frons and frontal vitta almost parallel. Fronto-orbital plate with 3-5 proclinate and 1-2 reclinate orbital bristles, and with some additional hairs at vertex. Parafacial plate bare. First and second antennomeres and base of third light brown to orange, remaining part of third antennomere blackish. Third antennomere 1.5-2.0× as long as second. Arista thickened in proximal half. Palpi yellow. Proboscis long. Thorax and legs grey pollinose. Fore tarsus with 2 elongate hairs on apicodorsal surface of tarsomeres 1-3, with hair-length decreasing in distal direction (Fig. 28). Abdomen ash-grey with a blackish grey median stripe. Colour pattern slightly changing with the incidence of light. T3 with a row of long, somewhat slender marginal bristles, T4-T5 with strong marginals.

♀. Like the male but with unmodified fore tarsus.

Length ♂♀. 6.5-9.5 mm.

Distribution. In Fennoscandia only from the USSR: Vib. – Western Europe to East Siberia, India (Kashmir), and Mongolia; south to North Africa. Not in the British Isles.

Biology: Unknown.

7. *Miltogramma villeneuvei* Verves, 1982
   Figs 27, 33, 52, 53.

*Miltogramma meigeni* Villeneuve, 1922, Ann. Sci. nat., Zool., 10(5): 342. Preocc. by Robineau-Desvoidy, 1863.
*Miltogramma villeneuvei* Verves, 1982a, Ént. Obozr., 61(1): 189. New name for *Miltogramma meigeni* Villeneuve, 1922.

Grey species. Antennae orange to blackish orange. Male fore tarsus with 1-2 erect hairs on tarsomeres 1-4. Abdomen grey to olive-grey pollinose.

♂. Head yellow. Frons and frontal vitta parallel. Fronto-orbital plate with 3-5 proclinate and 1-2 reclinate orbital bristles and with some additional hairs at vertex. Parafacial plate bare. Antennae orange to blackish orange, third antennomere often darker

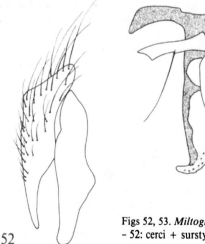

53

Figs 52, 53. *Miltogramma villeneuvei* Verves. – 52: cerci + surstyli, lateral view; 53: aedeagus.

52

37

than second. Third antennomere 1.8-2.1 × as long as second. Arista thickened in proximal 0.3-0.5. Palpi yellow. Proboscis short. Thorax and legs grey pollinose. Fore tarsus with 1-2 long hairs on apical part of ad and a surfaces of tarsomeres 1-4 (Fig. 27). Abdomen grey to olive-grey pollinose with an indication of a median stripe. Colour pattern not (or very slightly) changing with the incidence of light. T3 with bristly marginal setae, T4-T5 with a more distinct row of marginal bristles. Terminalia similar to those of *M. brevipilum,* but without the hump at base of surstyli (Fig. 52).

♀. Like the male but with unmodified fore tarsus.

Length ♂♀. 7.5-8.5 mm.

Distribution. Not recorded from Denmark. In Fennoscandia only from Finland: Sa. Verves (1986) lists *M. villeneuvei* from Sweden, but this is probably based on Ringdahl's (1937) record of *M. brevipilum,* as Verves does not give Sweden under the distribution of this species. – Palaearctic, from western Europe and North Africa to the Far East. Not in the British Isles.

Biology. Unknown.


# Genus *Pterella* Robineau-Desvoidy, 1863

*Pterella* Robineau-Desvoidy, 1863, Hist. nat. Dipt. Paris, 2: 121.
Type species: *Miltogramma grisea* Meigen, 1824.
*Setulia* Robineau-Desvoidy, 1863, Hist. nat. Dipt. Paris, 2: 124.
Type species: *Setulia cerceridis* Robineau-Desvoidy, 1863, = *Miltogramma grisea* Meigen, 1824.

Medium-sized flies of Old World distribution. Head generally large, with well-developed vibrissae, but otherwise often rather *Miltogramma*-like. Numerous frontals, often 2 proclinate and 1 reclinate orbital bristles, vertex rather densely haired. Parafacial plates bare or with very fine hairs, genae with hyaline hairs. Mid tibia with a single strong and often some weak ad bristles. Wings more or less shining due to reduced clothing-hairs, cell $r_{4+5}$ broadly open. Abdominal T5 often ventrally with hairs directed backwards and somewhat upwards.


## 8. *Pterella grisea* (Meigen, 1824)
Figs 54-58.

*Miltogramma grisea* Meigen, 1824, Syst. Beschr., 4: 230.

A medium-sized grey species with well-developed vibrissae and 2 proclinate orbitals. Abdomen grey with small spots, changing in colour with the incidence of light. Very like *Senotainia tricuspis,* but with shorter claws and numerous additional notopleural hairs.

♂. Head yellow with sparse whitish pollinosity. Frontal vitta broad, parallel. Fronto-orbital plate with 2 proclinate and 1 reclinate orbital bristles and several long hairs at vertex. Parafacial plate with fine hyaline hairs. Antennae more or less orange with a black arista. Third antennomere 2.4-2.8× as long as second. Arista thickened in proximal 0.6-0.7. Vibrissae as long as arista and situated well above lower margin of facial plate. No supravibrissal setae. Palpi yellow. Proboscis brown, of medium length. Thorax and legs black with grey pollinosity. Claws and pulvilli 0.75-0.80× as long as fifth tarsomere. Abdomen densely grey pollinose. Tergites with light brown hind margin and 3 small spots, the median spot of T3-T4 almost divided in two. Pattern distinct-

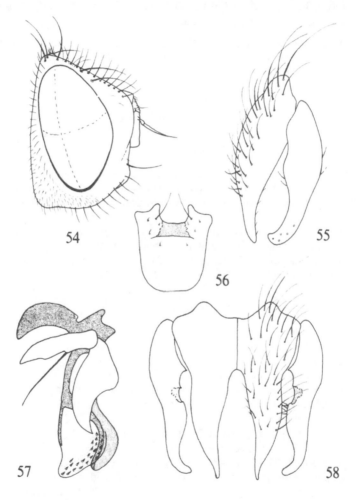

Figs 54-58. *Pterella grisea* (Meigen). – 54: head ♂; 55: cerci + surstyli, lateral view; 56: ST5 ♂; 57: aedeagus; 58: cerci + surstyli, posterior view.

ly changing with the incidence of light. T3 with a pair of median marginal bristles. T4 with a row of marginals. Terminalia *Miltogramma*-like (Figs 55-58), with a well-developed epiphallus.

♀. Like the male, with a slightly narrower frons and stronger marginals on abdominal T5.

Length ♂♀. 6.5-8.0 mm.

Distribution. Only few specimens known from Fennoscandia and Denmark. Denmark: F, B; Sweden: Öl.; and Finland: N. Not recorded from Norway. – Western Europe, east to East Siberia and Mongolia.

Biology. Recorded from nests of Sphecidae: *Cerceris*. A record from *Dociostaurus maroccanus* (Thunberg) (Orthoptera: Acrididae) in Spain (Thompson 1951) may actually refer to the prey of a sphecid wasp.

## Genus *Senotainia* Macquart, 1846

*Senotainia* Macquart, 1846, Mém. Soc. Sci. Agric. Lille, 1844: 295.
   Type species: *Senotainia rubriventris* Macquart, 1846.

The genus *Senotainia* is large and has representatives in all zoogeographical regions. The genus is in need of a thorough revision as it is rather ill-defined at present and contains several species which are included as they do not fit into any other genus. The Afrotropical fauna in particular is very diverse and rich in species, and many more await description. Characters generally associated with *Senotainia* are: frons not protruding, parafacial plate rather broad, and proboscis of medium length to long. Wings hyaline, cell $r_{4+5}$ open. Male claws and pulvilli long, abdomen often with somewhat protruding terminalia.

### Key to species of *Senotainia*

1   Palpi blackish brown .............................. 9. *conica* (Fallén)
–   Palpi yellow or reddish yellow ...................................... 2
2(1) Proboscis long, prementum 2.0-2.5×length of palpi. Head
     yellow with grey pollinosity. Parafacial plate bare ...... 12. *tricuspis* (Meigen)
–   Proboscis of medium length, prementum at most 1.7×as long
     as palpi. Head black with silvery grey pollinosity. Parafacial
     plate setulose ...................................................... 3
3(2) Arista thickened in proximal 0.5-0.6. Vibrissa situated slight-
     ly above level of lower eye-margin (Fig. 60). ♂: third an-
     tennomere 2.5-2.8×as long as second. Gonopods narrow, not
     serrated on dorsal margin (Fig. 71). ♀: vibrissal angle as pro-
     minent as frons .......................... 10. *puncticornis* (Zetterstedt)
–   Arista thickened in proximal 0.3-0.4. Vibrissa situated dis-
     tinctly above level of lower eye-margin (Fig. 61). ♂: third

40

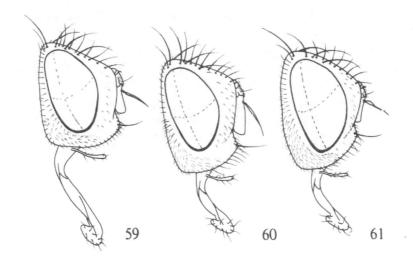

Figs 59-61. *Senotainia* spp., head profile ♂. – 59: *S. conica* (Fallén); 60: *S. puncticornis* (Zetterstedt); 61: *S. albifrons* (Rondani).

antennomere 1.6-2.0×as long as second. Gonopods broad, serrated on dorsal margin (Fig. 72). ♀: vibrissal angle slightly less prominent than frons . . . . . . . . . . . . . . . . . . . . . 11. *albifrons* (Rondani)

### 9. *Senotainia conica* (Fallén, 1810)
Figs 59, 62-65.

*Tachina conica* Fallén, 1810, K. VetenskAkad. Handl., (2) 31: 270.

Small species with light head and a greyish brown abdomen with ill-defined lateral black spots. Easily recognised by the brown palpi.

♂. Head silvery grey pollinose with a slightly brownish tinge at vertex and upper part of frontal vitta. Fronto-orbital plate with 2 proclinate and 1 reclinate orbital bristles and a few very short hairs on anterior half. Parafacial plate very broad, with scattered hairs. Vibrissal angle about as prominent as frons (Fig. 59). Antennae black, often slightly reddish at apical margin of second antennomere. Third antennomere about 1.5×as long as second. Arista thickened in proximal 0.3. Palpi dark brown. Proboscis with prementum 2.0-2.5×as long as palpi. Thorax grey pollinose, more olive-brown dorsally. Legs black, claws and pulvilli longer than fifth tarsomere. Abdomen mottled brown and black, tergites with ill-defined black spots laterally and often with a dark brown spot medially. T3-T4 with a pair of median marginal bristles. T5 with a row of marginals. Terminalia robust and protruding. Aedeagus without epiphallus, with a paired sclerite just below acrophallus (Fig. 64).

♀. Like the male but with head slightly paler, parafacial plate bare or with very few hairs. Claws and pulvilli short.

Length ♂♀. 3.5-5.5 mm.

Distribution. Common in Denmark and Fennoscandia about as far north as the Arctic Circle. – Palaearctic, from the British Isles to Mongolia.

Biology. Recorded from nests of Sphecidae: *Bembix, Crabro, Oxybelus, Philantus, Sphex, Tachytes;* Apidae: *Halictus.* Bred from *Crabro scutellatus* (Scheven), *Oxybelus uniglumis* (Linnaeus), and *Halictus gracilis* Morawitz in Finland (Tiensuu 1939; Nuorteva 1946).

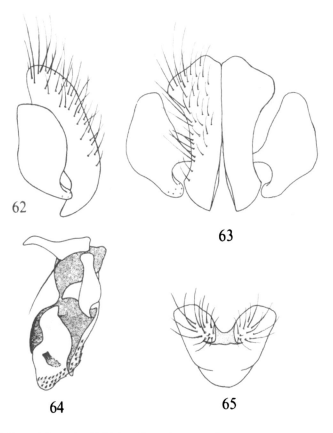

Figs 62-65. *Senotainia conica* (Fallén). – 62: cerci + surstyli, lateral view; 63: cerci + surstyli, posterior view; 64: aedeagus; 65: ST5 ♂.

10. *Senotainia puncticornis* (Zetterstedt, 1859)
   Figs 60, 67-71.

*Tachina imberbis* Zetterstedt, 1838, Insecta Lapp.: 636. Preocc. by Wiedemann, 1830.
*Miltogramma puncticornis* Zetterstedt, 1859, Dipt. Scand., 13: 6149.
*Ptychoneura crabronum* Kramer, 1920, Zool. Jb. Syst., 43: 329.

Rather variable in size, but often somewhat larger than *S. conica*. Palpi yellow and proboscis of medium length. Abdominal tergites with 3 brown spots.

 ♂. Fronto-orbital and parafacial plate silvery grey pollinose; frontal vitta black in dorsal view, more silvery grey in anterior view. Two proclinate and 1 reclinate orbital bristles. Fronto-orbital and parafacial plate with scattered hairs in addition to the bristles. Vibrissal angle about as prominent as frons. Antennae black, second antennomere with reddish apical margin. Third antennomere 2.5-2.8×length of second. Arista thickened in proximal 0.5-0.6, second aristomere longer than greatest diameter. Palpi yellow or orange. Proboscis of medium length (Fig. 60). Thorax grey pollinose, dorsally with olive-brown stripes. Legs black, claws and pulvilli as long as or slightly longer than fifth tarsomere. Abdomen grey with 3 spots on each tergite; the median spot lighter brown, lateral spots dark brown to black. T3 with a pair of median marginal bristles, T4-T5 with a row of marginals. Terminalia not enlarged, but often visible in lateral view. Aedeagus with a distinct hump on dorsal part of distiphallus, epiphallus reduced (Fig. 70).

 ♀. Similar to male but with parafacial hairs reduced and third antennomere only 1.5×as long as second. Abdominal pattern somewhat reduced.
   Length ♂♀. 4.0-6.5 mm.

Distribution. Rare in Denmark; rather common in Fennoscandia where it reaches as far north as MR in Norway, T. Lpm. in Sweden, and Li in Finland. – Central and northern Europe, East Siberia. Not in the British Isles.

Biology. Recorded from nests of Sphecidae: *Crossocerus.*

Note: The identity of *S. puncticornis* was revised by Pape (1986).


11. *Senotainia albifrons* (Rondani, 1859)
   Figs 61, 72.

*Sphixapata albifrons* Rondani, 1859, Dipt. Ital. Prodromus, 3: 225.

Very similar to *S. puncticornis,* especially in the female sex, but generally larger, although the size is very variable and of no value for identification purposes. Arista thickened in less than proximal half, and vibrissae situated slightly above lower eye-margin (Fig. 61).

 ♂. Colour as in *S. puncticornis*. Antennae short, third antennomere 1.6-2.0×as long as second. Copulatory apparatus very like that of *S. puncticornis,* but the gonopods seem to be characteristic: broad and with a serrate dorsal margin (Fig. 72). The shape of the gonopods is, however, somewhat variable.

♀. Colour variable, but often more extensively grey than in the male, and abdominal spots and mottles sometimes almost absent. The only reliable means of separating females of the present species from females of *S. puncticornis* is the shape of arista and the slight difference in head profile as given in the key.

Length ♂♀. 4.3-8.5 mm.

Distribution. Not recorded from Denmark. The only reliable record from Fennoscandia is a single male from Sweden: Sk., collected by Frey (ZMH). Ringdahl (1945:

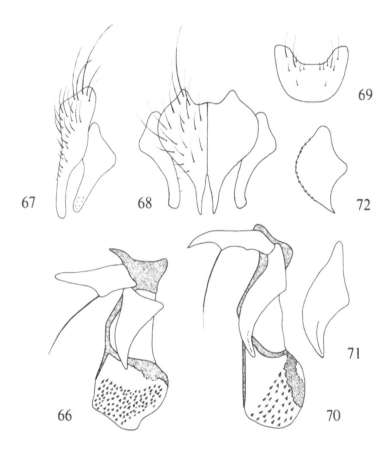

Fig. 66: *Senotainia tricuspis* (Meigen), aedeagus.
Figs 67-71. *Senotainia puncticornis* (Zetterstedt). – 67: cerci + surstyli, lateral view; 68: cerci + surstyli, posterior view; 69: ST5 ♂; 70: aedeagus; 71: gonopod.
Fig. 72. *Senotainia albifrons* (Rondani), gonopod.

35) notes that *Miltogramma imberbis* var. b of Zetterstedt (1859: 6149) is the present species, but I have not studied these specimens, which are probably in RMS (Roth Coll.). The record from Finland (Hackman 1980) is a misidentification of three females of *Actia crassicornis* (Meigen) (Tachinidae). – Widely distributed in southern and central parts of the Palaearctic region. Not in the British Isles. Recorded from the Afrotropical and Oriental regions and probably widespread.

Biology. Recorded from nests of Sphecidae: *Philantus, Prionyx, Sphex.*

12. *Senotainia tricuspis* (Meigen, 1838)
    Fig. 66.

*Miltogramma tricuspis* Meigen, 1838, Syst. Beschr., 7: 234.

A medium-sized species, generally larger than *S. puncticornis.* Easily recognised by the yellowish head and long proboscis.

♂. Head silvery grey pollinose and yellow except for blackish vertex, gena, postgena, and occiput. Fronto-orbital plate with 2 proclinate and 1 reclinate orbital bristles and a few short hairs on anterior part. Parafacial plate bare. Vibrissal angle about as prominent as frons. Antennae black with reddish apical margin of second antennomere. Third antennomere about 1.5 × as long as second. Arista thickened in proximal 0.4. Palpi yellow. Proboscis with prementum 2.0-2.5 × as long as palpi. Thorax and legs black with grey pollinosity. Claws and pulvilli longer than fifth tarsomere. Abdomen with grey pollinosity and 3 black spots on each of T3-T5. A pair of median marginal bristles on T3 (often weak). Terminalia: aedeagus similar to that of *S. puncticornis,* but more robust and with a shorter dorsal plate (Fig. 66).

♀. Like the male but with claws and pulvilli about as long as fifth tarsomere. Abdomen more extensively greyish pollinose and tergal spots reduced.

Length ♂♀. 6.0-9.0 mm.

Distribution. Not in Denmark, and only a single record from Fennoscandia: a male from Sweden: Sk. (ZML). – Palaearctic, from western Europe and North Africa to Mongolia.

Biology. Recorded from nests of Sphecidae: *Ectemnius, Philantus.* A parasitoid on bumblebees and the honey bee *Apis mellifera* Linnaeus (Boiko 1939, 1948).

## Genus *Amobia* Robineau-Desvoidy, 1830

*Amobia* Robineau-Desvoidy, 1830, Essai Myod.: 96.
  Type species: *Amobia conica* Robineau-Desvoidy, 1830, = *Tachina signata* Meigen, 1824.
*Pachyophthalmus* Brauer & Bergenstamm, 1889, Denkschr. Akad. Wiss. Wien, Kl. math.-naturw., 56: 117.
  Type species: *Tachina signata* Meigen, 1824.

A well-defined genus, represented in all zoogeographical regions. Head with a rounded profile. Frontal vitta parallel, orbital bristles hair-like and numerous, and antennal insertion below centre of eye. Antennae of medium length, arista long. Male tarsus rather robust, with claws and pulvilli as long as or longer than fifth tarsomere. Epiphallus absent.

The species are kleptoparasites with a wide range of hosts among solitary Vespidae, Sphecidae, and Apidae, and they attack both terrestrial and arboreal hosts.

## Key to species of *Amobia*

1 ♂: aedeagus widening apically into a very large phallotreme. Gonopods of medium length (Fig. 75). ♀: abdominal spots on T3-T4 well-defined, not coalescing at posterior margin. Uterus without ventral sclerotisation. Often two elongate and one oval spermathecae (Fig. 81) ............................. 13. *signata* (Meigen)
- ♂: phallotreme not greatly enlarged. Gonopods long and slender (Fig. 80). ♀: abdominal spots on T3-T4 tending to coalesce at posterior margin. Uterus with a ventral sclerotisation (Fig. 78). Often two oval and one spherical spermathecae (Fig. 79). ................................. 14. *oculata* (Zetterstedt)

### 13. *Amobia signata* (Meigen, 1824)
Figs 73-77, 81.

*Tachina signata* Meigen, 1824, Syst. Beschr., 4: 303.

Dark grey with 3 distinct stripes on thorax, and abdominal tergites with distinct spots. Rather variable in size.

♂. Head with a rounded profile, antennae inserted just below centre of eye (Fig. 77). Fronto-orbital and parafacial plates grey pollinose. Frontal vitta black, parallel. Proclinate orbital bristles numerous and hair-like, only slightly stronger than adjacent fronto-orbital hairs. Parafacial plate with long hairs on upper 0.5-0.7. Antennae black, third antennomere about 1.5 × length of second. Arista much longer than third antennomere, thickened in proximal 0.3. Second aristomere longer than greatest aristal diameter. Vibrissae well-developed, genal groove densely haired. Palpi black. Proboscis dark brown, of medium length. Thorax grey pollinose with 3 distinct dark brown stripes. Legs black, grey pollinose; tarsus rather robust with claws and pulvilli about as long as fifth tarsomere. Wings with a blackish brown basicosta. Abdomen densely grey pollinose with 3 spots on all tergites, which more or less coalesce on T5. T1 + 2 with 1-2 pairs of median marginal bristles, T3 with at least 2 pairs of marginals, and T4-T5 with a distinct row of marginals of which the median pair has advanced to a more anterior position. Terminalia: T6 and syntergosternite 7 + 8 fused, but with a distinct suture marking the segmental origin. Aedeagus large, with a very broad phalotreme (Fig. 75). Epiphallus absent.

♀. Like the male in general appearance. Inner eye-facets distinctly enlarged, parafacial hairing slightly shorter and sparser, claws and pulvilli shorter than fifth tarsomere.

Length ♂♀ . 5.0-8.5 mm.

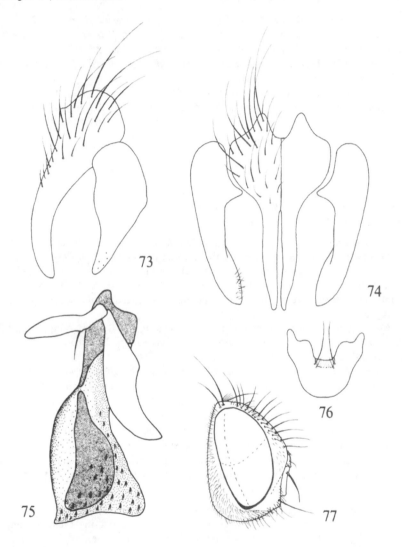

Figs 73-77. *Amobia signata* (Meigen). – 73: cerci + surstyli, lateral view; 74: cerci + surstyli, posterior view; 75: aedeagus; 76: ST5 ♂; 77: head profile ♂.

Distribution. Common in Denmark and southern Sweden: Sk., Upl. Not recorded from Norway or Finland. – Widely distributed in the Palaearctic and Afrotropical regions. Records from the Nearctic region refer to the closely-related species (or subspecies) *A. aurifrons* (Townsend).

Biology. Recorded from nests of Vespidae: *Ancistrocerus, Eumenes, Odynerus, Discoelius, Synagris;* Sphecidae: *Cerceris, Crossocerus, Ectemnius, Pemphredon, Psen, Psenulus, Sceliphron, Trypoxylon;* Apidae: *Andrena, Megachile, Osmia.*

Théodorides (1954) records *A. signata* as a parasite of the acridid grasshopper *Aeolopus strepens* (Latreille).

Bred from nests of *Odynerus reniformis* (Gmelin), *O. spinipes* (Linnaeus), and *Megachile rotundata* (Fabricius) in Denmark (Lundbeck 1927; unpublished); Weis (1960) tentatively mentions *Ancistrocerus parietinus* (Linnaeus) as host.

14. *Amobia oculata* (Zetterstedt, 1844)
   Figs 78-80.

*Miltogramma oculata* Zetterstedt, 1844, Dipt. Scand., 3: 1212.
*Pachyophthalmus distortus* Allen, 1926, Proc. U.S. natn. Mus., 68: 15.

Extremely similar to *A. signata* in both sexes. Males are quite easily separated by the aedeagus and gonopods, but the females are very difficult to identify with certainty. There seems to be a slight difference in the abdominal coloration, that of *A. signata* consisting of well-defined spots which are clearly separated on T3-T4, while these spots are slightly more triangular in *A. oculata* and tend to coalesce at the posterior margin.

Males of *A. oculata* may be recognised by the long and slender gonopods and the narrow phallotreme (Fig. 80) . *Amobia signata* has gonopods of medium length and a markedly expanded phallotreme (Fig. 75).

A key to females is given by Draber-Mońko (1966), who gives the width of a fronto-orbital plate relative to the width of the frontal vitta as a diagnostic character (< 1 for *A. oculata*, ≧ 1 for *A. signata*). While this may hold for Nearctic specimens of *A. oculata* when compared with Palaearctic specimens of *A. signata,* I have not found that it works with any degree of certainty for Fennoscandian and Danish specimens. The shape of spermathecae may be of importance in species recognition, as given in the key, and another character may be the presence or absence of a ventral sclerotisation in the uterus (Fig. 78). I have found this sclerotisation in all dissected females of *A. oculata* but not in *A. signata.*

Distribution. No records from Denmark. The species seems to be common in southern Norway and in Sweden north of Sk. From Finland a single female from St. – A widely distributed Holarctic species. Not in the British Isles.

Biology. Bred from nests of Sphecidae: *Trypoxylon;* Vespidae: *Antherhyncium, Eumenes, Rhynchium, Stenodynerus, Symmorphus* (Kurahashi 1974).

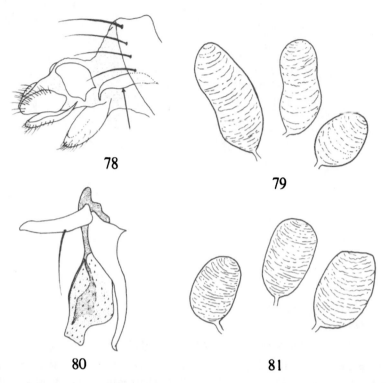

Figs 78-80. *Amobia oculata* (Zetterstedt). – 78: terminalia ♀; 79: spermathecae; 80: aedeagus.
Fig. 81. *Amobia signata* (Meigen), spermathecae.

## Genus *Phylloteles* Loew, 1844

*Phylloteles* Loew, 1844, Stettin. ent. Ztg, 5: 168.
    Type species: *Phylloteles pictipennis* Loew, 1844.

Characterised by the flattened arista, which is more or less leaflike in the male sex. Frons slightly protruding, antennae short. Orbital bristles 3-5, parafacial plate bare or with fine, white hairs, genal hairs often white. Sexual dimorphism may be well-developed, and the male may possess spotted or fumose wing-tips, and the vibrissae and genal bristles may be completely reduced.

    The genus is known in several species from the Afrotropical region, one of these being known as a predator of locust egg-pods. A few species occur in the Palaearctic and Oriental regions.

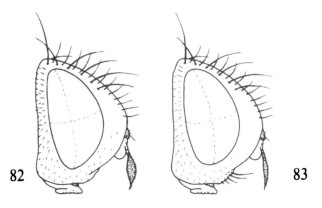

Figs 82, 83. *Phylloteles pictipennis* Loew, head profile. – 82: ♂; 83: ♀.

### 15. *Phylloteles pictipennis* Loew, 1844
Figs 82,83.

*Phylloteles pictipennis* Loew, 1844, Stettin. ent. Ztg, 5: 168.

Grey species with short antennae and flattened arista. Abdominal tergites with 3 black spots. Sexual dimorphism well-developed, males with leaf-like arista and spotted wings, and without vibrissae.

♂. Head silvery pollinose, frontal vitta often with a yellowish tinge. Fronto-orbital plate with 3-4 proclinate and 1 reclinate orbital bristles and a few dark hairs at vertex. Lower part of fronto-orbital plate and all of parafacial plate and gena covered with fine white hairs. No vibrissae or genal bristles. Antennae short, third antennomere 1.5-2.0 × as long as second. Arista leaf-like flattened and white or yellowish white in distal half. Palpi yellow. Proboscis short. Thorax grey pollinose. Wings with infuscated spots apically. Abdomen grey pollinose and with 3 black dorsal spots on each of T3-T5.

♀. Yellowish tinge of frontal vitta more pronounced. White hairs of fronto-orbital and parafacial plate reduced. Arista only slightly flattened (Fig. 83). Vibrissae present, only slightly stronger than adjacent genal bristles.

Length ♂♀. 4.0-6.5 mm.

Distribution. In Fennoscandia only from USSR: Vib. The record from Denmark in Verves (1986) needs confirmation. – Central and southern Europe, east to West Siberia and Turkey.

Biology. Unknown.

# Genus *Oebalia* Robineau-Desvoidy, 1863

*Oebalia* Robineau-Desvoidy, 1863, Hist. nat. Dipt. Paris, 2: 414.
Type species: *Oebalia anacantha* Robineau-Desvoidy, 1863, = *Tachina cylindrica* Fallén, 1810.
*Ptychoneura* Brauer & Bergenstamm, 1889, Denkschr. Akad. Wiss. Wien, Kl. math.-naturw., 56: 104.
Type species: *Tachina rufitarsis* Meigen, 1824, = *Tachina minuta* Fallén, 1810.

Small to medium-sized black flies with grey pollinosity. Antennae medium-sized to long, and arista characteristically thickened in about proximal 0.75. Legs rather robust, with broad tarsomeres, and claws and pulvilli long. Abdomen oval with terminalia somewhat protruding, often with 3 black spots on T3-T4, but the spots may coalesce into a broad continuous black hind-margin. Median marginal bristles indistinct.

Male cerci sickle-shaped and bent backwards, aedeagus with the dorsal plate prolonged into a bilobed extension.

Species of *Oebalia* are ovo-larviparous kleptoparasites, especially associated with stalk-nesting Sphecidae (Tolstova 1962; Day & Smith 1980). The genus is represented in the Palaearctic region by 8 species. The few New World species are in need of revision.

## Key to species of *Oebalia*

1  Tarsi yellowish. Facial ridge with 4-7 bristles (Fig. 84). Abdomen sparsely pollinose, with ill-defined spots . . . . . . . . . 16. *minuta* (Fallén)
–  Tarsi black. Facial ridge with short hairs on lower part. Abdomen with distinct spots . . . . . . . . . . . . . . . . . . . . . . . . . . . . . . . . . . . . . . 2

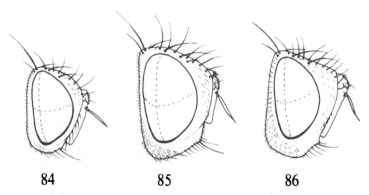

84      85      86

Figs 84-86. *Oebalia* spp., head profile ♂. – 84: *O. minuta* (Fallén); 85: *O. cylindrica* (Fallén); 86: *O. sachtlebeni* Rohdendorf.

2(1)  Abdominal spots coalescing on hind-margin. Parafacial plate haired on upper 0.5-0.7, with narrowest width equal to
about the length of fifth tarsomere of fore tarsus (Fig. 85).
Third antennomere 2.5-3.0 × as long as second in both sexes
......................................... 17. *cylindrica* (Fallén)
-  Abdominal spots separated by pollinosity which reaches the
hind-margin. Parafacial plate haired on its entire length,
with narrowest width 1.2-1.5 × the length of fifth tarsomere
of fore tarsus (Fig. 86). Third antennomere 3.0-4.0 (♀) or
4.5-5.5 (♂) × as long as second .............. 18. *sachtlebeni* Rohdendorf

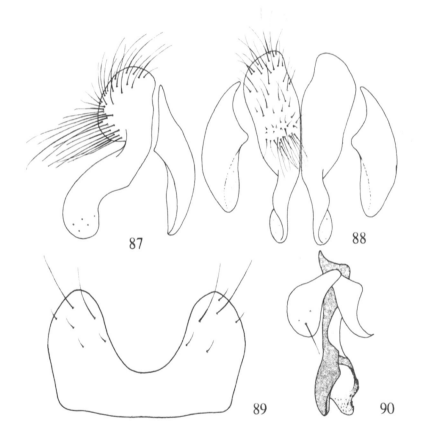

Figs 87-90. *Oebalia minuta* (Fallén). - 87: cerci + surstyli, lateral view; 88: cerci + surstyli,
posterior view; 89: ST5 ♂; 90: aedeagus.

16. *Oebalia minuta* (Fallén, 1810)
   Figs 84, 87-90.

*Tachina minuta* Fallén, 1810, K. VetenskAkad. Handl., (2) 31: 275.
*Tachina rufitarsis* Meigen, 1824, Syst. Beschr., 4: 410.

Black species with yellowish tarsi. Facial ridges with a row of bristles; genae and parafacial plate narrow.

♂. Fronto-orbital plate sparsely grey dusted, with an olive-brown tinge at vertex. Frontal vitta black. Parafacial plate silvery grey pollinose. Two proclinate and 1 reclinate orbital bristles. Lower half of fronto-orbital plate with some black hairs, parafacial plate bare except for a few hairs in uppermost part. Narrowest width of parafacial plate (in profile) narrower than length of fifth tarsomere of fore tarsus; gena narrow. Facial ridge with 4-7 bristles. Third antennomere 4.0-5.0× as long as second. Arista slightly shorter than third antennomere, thickened in proximal 0.75. Proboscis short. Palpi and proboscis dark brown basally, lighter brown to yellowish apically. Thorax sparsely grey pollinose, legs black with yellowish tarsi. Tarsomeres broadening apically, claws about as long as fifth tarsomere. Abdomen short and oval. T1 + 2 black, without pollinosity; T3-T4 with grey pollinosity on anterior, especially antero-lateral, margins. T5 with grey pollinosity on anterior 0.5. Terminalia rather robust and somewhat protruding. Cerci distinctly bent and with a basal tuft of hairs directed posteriorly (Fig. 87). Aedeagus strikingly different from other species of *Oebalia,* with the two-lobed extension of the dorsal plate broad and hood-like (Fig. 90).

♀. As male but with more extensive pollinosity on abdomen. Terminalia distinctly protruding, T6-T7 densely setose on hind-margin.

Length ♂♀. 3.5-5.5 mm.

Distribution. Common in Denmark and southern Sweden. A few records from southern Finland and southernmost Norway. – Widely distributed in the Palaearctic region, from the British Isles to Japan. Probably in the Nearctic region as well.

Biology. Recorded from nests of Sphecidae: *Crossocerus, Rhopalum.*


17. *Oebalia cylindrica* (Fallén, 1810)
   Figs 85, 91-95.

*Tachina cylindrica* Fallén, 1810, K. VetenskAkad. Handl., (2)31: 279.

Blackish species. Antennae of medium length, third antennomere shorter than arista. Abdomen with grey pollinosity tinged with olive, and large black spots on T3-T4.

♂. Fronto-orbital plate silvery grey pollinose with an olive tinge. Frontal vitta black. Parafacial plate silvery grey pollinose. Fronto-orbital plate with 2 proclinate and 1 reclinate orbital bristles and some black hairs. Parafacial plate with some hairs on inner margin of upper 0.5-0.7. Facial ridge with black hairs on about lower 0.5. Narrowest width of parafacial plate (in profile) equal to the length of fifth tarsomere of

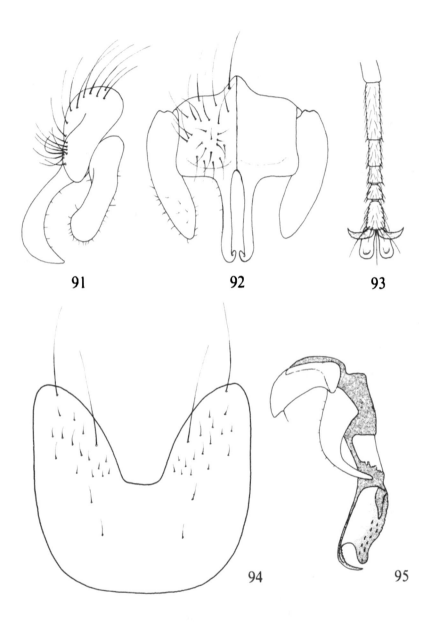

Figs 91-95. *Oebalia cylindrica* (Fallén). – 91: cerci + surstyli, lateral view; 92: cerci + surstyli, posterior view; 93: right fore tarsus ♂, dorsal view; 94: ST5 ♂; 95: aedeagus.

fore tarsus; gena moderately narrow. Third antennomere 2.5-3.0×as long as second. Proboscis short. Palpi and proboscis brown. Thorax sparsely grey pollinose. Legs black, with tarsi rather robust. Claws about as long as fifth tarsomere. Abdomen oval. T1+2 black and without pollinosity. T3-T4 with silvery grey, often somewhat olive-tinged pollinosity and 3 black, more or less rectangular spots which coalesce at hind-margin. T5 silvery grey pollinose with only a narrow median black stripe. Terminalia robust and somewhat protruding. Cerci distinctly bent and hook-shaped. Aedeagus with the dorsal plate extended into two curled processes (Fig. 95).

♀. Like the male, but generally a little larger. Terminalia distinctly protruding, T6-T7 with marginal bristles.

Length ♂♀. 3.0-6.0 mm.

Distribution. Not rare in Denmark and Fennoscandia, reaching north of the Arctic Circle in Sweden. – Widely distributed in western Europe, East Siberia, and Mongolia.

Biology. Recorded from nests of Sphecidae: *Crossocerus, Ectemnius, Rhopalum, Trypoxylon.*

Bred from *Crossocerus annulipes* (Lepeletier & Brullé), *C. cinxius* (Dahlbom), *C. capitosus* (Schuckard), *Rhopalum coarctacum* (Scopoli), and *Trypoxylon attenuatum* Smith in Finland (Tiensuu 1939). In Denmark bred from *Crossocerus* sp. and *Ectemnius cephalotes* (Olivier) (Lundbeck 1927).


18. *Oebalia sachtlebeni* Rohdendorf, 1963
    Figs 86, 96-99.

*Oebalia sachtlebeni* Rohdendorf, 1963b, Beitr. Ent., 13: 448.

A grey species with a somewhat large head. Males with third antennomere longer than arista. Abdomen with well-defined black spots.

♂. Fronto-orbital and parafacial plates silvery grey pollinose, slightly sparser at vertex. Frontal vitta black. Two proclinate (occasionally 3) and 1 reclinate orbital bristles. Fronto-orbital plate with a few hairs in addition to the bristles, parafacial plate rather densely haired. Facial ridge haired on lower 0.4. Narrowest width of parafacial plate about 1.2-1.5×the length of fifth tarsomere of fore tarsus. Gena moderately broad. Third antennomere long, 4.5-5.5×as long as second and slightly longer than arista. Arista thickened in proximal 0.8. Thorax greyish pollinose. Legs black, tarsomeres broadening apically, claws about as long as fifth tarsomere. Abdomen grey pollinose. T1+2 with pollinosity only laterally; T3-T4 with the pollinosity leaving 3 distinct triangular black spots, not coalescing at hind-margin; T5 with a more or less distinct median stripe. Terminalia somewhat protruding, aedeagus like *O. cylindrica,* but cerci slightly stronger and more bent (Fig. 96).

♀. Like the male but parafacial hairs shorter and third antennomere only 3.4-4.0× as long as second. The short third antennomere gives the female a superficial similarity to the female *O. cylindrica,* but it is easily separated by the broader parafacial plate, the extension of the parafacial hairing to below apex of third antennomere, and the abdominal pollinosity reaching hind-margins of tergites between the spots.

Length ♂♀. 4.5-6.5 mm.

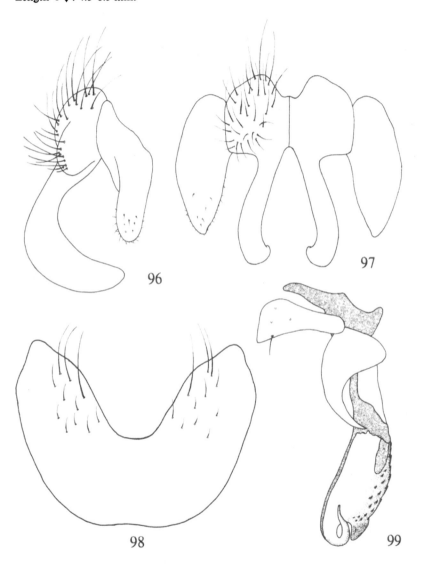

Figs 96-99. *Oebalia sachtlebeni* Rohdendorf. – 96: cerci + surstyli, lateral view; 97: cerci + surstyli, posterior view; 98: ST5 ♂; 99: aedeagus.

Distribution. Not rare in Denmark. In Fennoscandia north to Upl. in Sweden and Ø in Norway. Not recorded from Finland. – Europe east to western USSR. Not in the British Isles.

Biology. Recorded from nests of Sphecidae: *Pemphredon, Rhopalum* (Draber-Mońko 1978).

Note. Ringdahl's (1952) record of *O. praeclusa* (Pandellé, 1895) from Sweden: Sm. refers to the present species.

## Genus *Hilarella* Rondani, 1856

*Hilarella* Rondani, 1856, Dipt. Ital. Prodromus, 1: 70.
Type species: *Miltogramma hilarella* Zetterstedt, 1844.

Small grey or yellowish species. Head with slightly protruding frons in both sexes, parafacial plate haired, antennae of medium length, arista with short hairs. Wing cell $r_{4+5}$ closed at wing-margin or with a short petiole. Costal spine strong. Last section of $CuA_1$, from dm-cu to wing-margin, about $0.5 \times$ as long as previous section, from bm-cu to dm-cu. Claws and pulvilli short. Abdomen conically tapering, tergites with brown spots which may be rather reduced. A moderate sex-dimorphism is present; males with frons somewhat more protruding, parafacial hairing longer and denser, sensory hairs of fore tarsus elongated, and colour often slightly darker.

*Hilarella* is very reminiscent of *Taxigramma*, but is separated by the haired arista and the proportionally shorter distance from dm-cu to wing-margin. Five species, distributed in the Holarctic, Afrotropical, and northern Neotropical regions.

### Key to species of *Hilarella*

1 Grey species. Three (seldom 2) proclinate orbitals. Abdomen
with 3 well-developed spots . . . . . . . . . . . . . . . . . . . . . 19. *hilarella* (Zetterstedt)
– Yellowish species. Two proclinate orbitals (seldom 3). Abdominal pattern reduced, with small lateral and a pair of small
median spots . . . . . . . . . . . . . . . . . . . . . . . . . . . . . . . . 20. *stictica* (Meigen)

### 19. *Hilarella hilarella* (Zetterstedt, 1844)
Figs 100-104, 110, 117.

*Miltogramma hilarella* Zetterstedt, 1844, Dipt. Scand., 3: 1212.

A grey species with silvery pollinose head, 3 (seldom 2) proclinate orbital bristles, and a grey abdomen with distinct dark brown spots.

♂. Frons conically protruding. Fronto-orbital and parafacial plate silvery grey pollinose, frontal vitta with silvery grey pollinosity at vertex, more sparsely grey on anterior 0.75. Three pairs of strong proclinate orbital bristles, but occasionally only 2 present on one or both sides. Fronto-orbital plate with a few scattered hairs in addition to the bristles, parafacial plate densely haired (Fig. 110). First and second antennomeres reddish, third antennomere greyish black, often reddish at level of aristal insertion. Third antennomere 2.0-3.0× as long as second. Arista with distinct hairs, the longest of which are slightly longer than greatest aristal diameter. Palpi yellow. Proboscis of medium length, prementum + labellum as long as or slightly shorter than arista. Thorax black with grey pollinosity. Legs grey pollinose. Trochanters, tibiae, and femoro-tibial joints reddish, femora of all legs brownish black to black. Mid tibia with

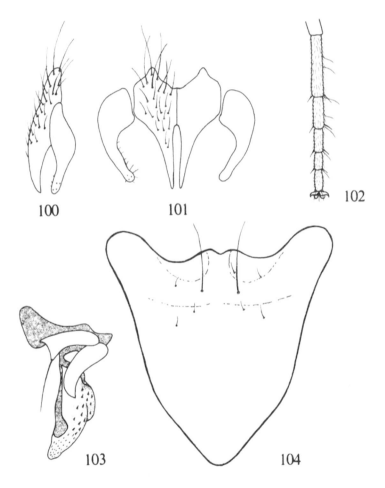

Figs 100-104. *Hilarella hilarella* (Zetterstedt). – 100: cerci + surstyli, lateral view; 101: cerci + surstyli, posterior view; 102: right fore tarsus ♂, dorsal view; 103: aedeagus; 104: ST5 ♂.

or without an ad bristle. Hind tibia often distinctly darker than fore and mid tibiae, but the extent of the red colour somewhat variable. Tarsomeres 2-4 of fore tarsus with elongated sensory hairs in anteroventral position (Fig. 102). Cell $r_{4+5}$ closed at wing-margin or short petiolate. Lower calypters infuscated in centre. Abdomen grey pollinose; in dorsal view with 3 dark brown spots on each of T3-T5, often coalescing on T5. A pair of median marginal bristles situated on the median spots of T3-T5. Terminalia: aedeagus with a robust epiphallus which is distinctly swollen apically (Fig. 103).

♀. Like the male, but with a more light brown tinge on upper part of frons, dorsal part of thorax, and abdomen. Hairing of parafacial plate much shorter and sparser, and sensory hairs of fore tarsus of normal length. Mid tibia always with an ad bristle.
Length ♂♀. 3.5-6.0 mm.

Distribution. In Denmark recorded from EJ and B. Reaching as far north as Lu. Lpm. in Sweden and Kb in Finland. Not recorded from Norway. – Europe except for the British Isles, east to Mongolia. Widely distributed in the Nearctic region and northern parts of the Neotropical region.

Biology. Recorded from nests of Sphecidae: *Ammophila*. Bred from a camel cricket (Orthoptera: Rhaphidophoridae) (Arnaud 1954).
Bred from a nest of *Ammophila sabulosa* (Linnaeus) in Finland (Tiensuu 1939).


## 20. *Hilarella stictica* (Meigen, 1824)
Figs 105-109, 111.

*Miltogramma stictica* Meigen, 1824, Syst. Beschr., 6: 367.
*Miltogramma siphonina* Zetterstedt, 1844, Dipt. Scand., 3: 1213.

Yellowish grey. Head with 2 proclinate orbital bristles and sparse parafacial hairing. Parafacial plate broad. Abdomen with reduced abdominal spots; almost unicolourous.

♂. Head with frons somewhat protruding (Fig. 111). Frontal vitta yellow, fronto-orbital and parafacial plates yellow with a light grey to almost whitish pollinosity. Two proclinate orbitals, seldom 3. Fronto-orbital plate without hairs in addition to the usual bristles. Parafacial plate with scattered black hairs, seldom densely haired. First and second antennomeres orange, third antennomere grey, more or less orange at base and 1.4-2.0× as long as second. Arista distinctly haired, longest hairs as long as or slightly longer than greatest aristal diameter. Palpi yellow. Proboscis of medium length, prementum + labellum slightly longer than arista. Thorax yellowish grey pollinose. Legs grey pollinose. Femora and tarsi blackish; trochanters, femoro-tibial joints, and tibiae orange. Mid tibia with 2 subequal v and 2 (-3) pd, occasionally with 1 ad. Tarsomeres 1-4 of fore tarsus with long sensory hairs in anteroventral position (Fig. 107). Cell $r_{4+5}$ closed at wing-margin or short petiolate. Lower calypters infuscated in centre. Abdomen yellowish grey, in dorsal view with a pair of small, brown lateral spots and a pair of small median spots on each of T3-T5. T1+2 with only lateral spots. T3-T4 each with a pair of marginal bristles situated on the median spots, T5 with a row of marginals. Terminalia similar to *H. hilarella* but with the epiphallus slightly less swollen apically (Fig. 108).

♀. Like the male, but with even sparser and shorter parafacial hairs. Mid tibia always with a strong ad bristle, sensory hairs of fore tarsus of normal length.

Length ♂♀. 4.0-6.0 mm.

Distribution. Common on dry grasslands, heaths, and warm sandy places in Denmark and southern Sweden. Not recorded from Norway or Finland. – Europe, North Africa, and western USSR, east to Mongolia. Not in the British Isles.

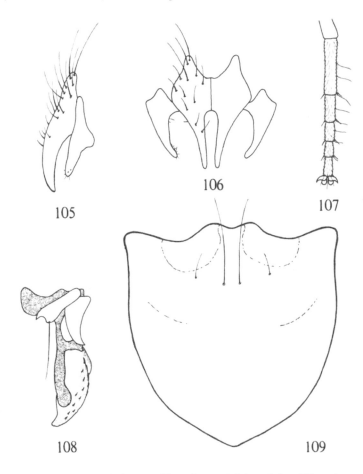

Figs 105-109. *Hilarella stictica* (Meigen). – 105: cerci + surstyli, lateral view; 106: cerci + surstyli, posterior view; 107: right fore tarsus ♂, dorsal view; 108: aedeagus; 109: ST5 ♂.

Biology. Recorded from nests of Sphecidae: *Ammophila, Bembix, Sphex.* Ferton (1901) and Maneval (1929) describe the oviposition.

Note. Verves (1986) treats *H. stictica* and *H. siphonina* as distinct species, but I have been unable to confirm this (Pape 1986).

## Genus *Taxigramma* Perris, 1852

*Taxigramme* Macquart, 1850:359. Vernacular name unavailable in nomenclature.
*Taxigramma* Perris, 1852, Annls Soc. linn. Lyon: 209.
   Type species: *Taxigramma pipiens* Perris, 1852, = *Miltogramma heteroneura* Meigen, 1830.
*Heteropterina* Macquart, 1854, Annls Soc. ent. Fr., (3)2: 426.
   Type species: *Miltogramma heteroneura* Meigen, 1830.
*Paragusia* Schiner, 1861, Wien. ent. Mschr., 5: 123.
   Type species: *Paragusia frivaldzkii* Schiner, 1862, = *Tachina elegantula* Zetterstedt, 1844.

Small species. Head with frons more or less protruding. Antennae of medium length or long, arista bare or micropubescent. Wing cell $r_{4+5}$ closed at wing-margin or

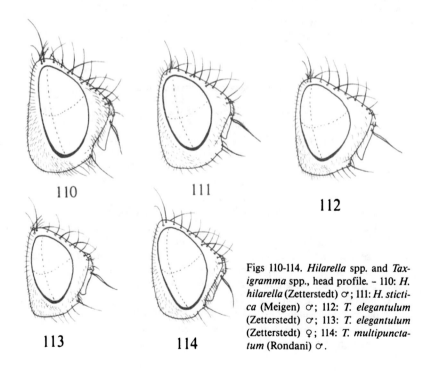

110   111

112

113   114

Figs 110-114. *Hilarella* spp. and *Taxigramma* spp., head profile. – 110: *H. hilarella* (Zetterstedt) ♂; 111: *H. stictica* (Meigen) ♂; 112: *T. elegantulum* (Zetterstedt) ♂; 113: *T. elegantulum* (Zetterstedt) ♀; 114: *T. multipunctatum* (Rondani) ♂.

short-petiolate. Costal spine strong. Vein CuA₁ with last section, from dm-cu to wing-margin, 1.0-2.0× as long as section from bm-cu to dm-cu. Abdomen conically tapering, tergites with 3 dorsal spots. A moderate sexual dimorphism present. Males often darker and with denser pollinosity, and with frons slightly more protruding. Male fore tarsi often with sensory hairs elongated, or with claws and pulvilli long (*T. heteroneurum*).

About 10 species, distributed in the Holarctic, Afrotropical, and Oriental regions.

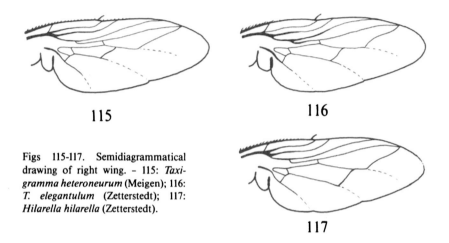

Figs 115-117. Semidiagrammatical drawing of right wing. – 115: *Taxigramma heteroneurum* (Meigen); 116: *T. elegantulum* (Zetterstedt); 117: *Hilarella hilarella* (Zetterstedt).

### Key to species of *Taxigramma*

1    Last section of CuA₁, from dm-cu to wing-margin, 2.0× length of section from bm-cu to dm-cu (Fig. 115). Parafacial plate bare or with a row of fine hairs on lower part. Proboscis of medium length. ♂: claws and pulvilli about as long as fifth tarsomere (Fig. 120) . . . . . . . . . . . . . . . . . . . . 21. *heteroneurum* (Meigen)
–    Last section of CuA₁ 1.0-1.4× length of previous section. Parafacial plate with distinct black setae. Proboscis short. Claws and pulvilli of both sexes distinctly shorter than fifth tarsomere . . . . . . . . . . . . . . . . . . . . . . . . . . . . . . . . . . . . . . . . . . . . . . . . 2
2(1)  Last section of CuA₁ 1.0× length of previous section. Parafacial setae rather scattered . . . . . . . . . . . . . . . . . . 22. *elegantulum* (Zetterstedt)
–    Last section of CuA₁ 1.4× length of previous section. Parafacial setae bristly and confined to one row at inner margin (Fig. 114). . . . . . . . . . . . . . . . . . . . . . . . . . . . . . . . [*multipunctatum* (Rondani)]
    (The record of Ringdahl (1937:33) from Öland is a specimen of *T. heteroneurum* misidentified by Villeneuve as *T. multipunctatum* (Rondani, 1859).)

62

## 21. *Taxigramma heteroneurum* (Meigen, 1830)
Figs 115, 118-122.

*Miltogramma heteroneura* Meigen, 1830, Syst. Beschr., 6: 367.

Brownish species. Wings with last section of CuA$_1$ about 2×as long as previous section. Abdominal tergites with 3 dark spots. Male claws and pulvilli long.

♂. Frons only slightly protruding. Fronto-orbital and parafacial plates silvery grey pollinose, frontal vitta brownish grey. Fronto-orbital plate with strong frontals and 2 proclinate orbitals, otherwise bare. Parafacial plate bare or with a single row of very fine hairs on lower part. First and second antennomeres reddish; third antennomere greyish black, more or less reddish at base, about 2×as long as second. Arista bare or micropubescent. Palpi yellow. Proboscis of medium length. Thorax grey pollinose.

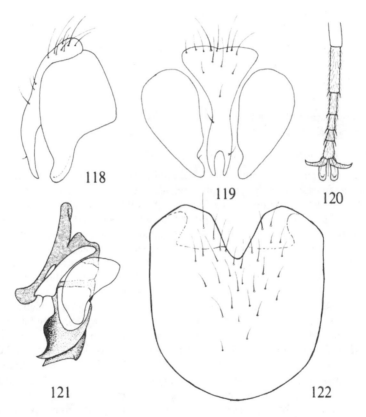

Figs 118-122. *Taxigramma heteroneurum* (Meigen). – 118: cerci + surstyli, lateral view; 119: cerci + surstyli, posterior view; 120: right fore tarsus ♂, dorsal view; 121: aedeagus; 122: ST5 ♂.

63

Legs dark brown to blackish brown, with light femoro-tibial joints. The colour of the legs is variable with tarsomeres broadening apically and claws and pulvilli long (Fig. 120). Claws and pulvilli of mid and hind legs about as long as fifth tarsomere. Wing cell $r_{4+5}$ closed at wing-margin, costa with a row of short bristles. Last section of $CuA_1$ 2× as long as previous section (Fig. 115). Abdomen brownish with light hind-margins. T1+2-T5 each with 3 black spots which may coalesce on T5. A pair of marginal bristles on T3-T4, a row of marginals on T5. Terminalia very characteristic, with cerci fused basally, and aedeagus unlike other species of *Taxigramma* (Figs 119, 121).

♀. Lighter than male. Antennae reddish except for arista, but occasionally with third antennomere darkened apically. Thorax brownish grey pollinose. Tarsi black, legs otherwise orange. Claws and pulvilli short.

Length ♂♀. 3.5-5.5 mm.

Distribution. Rare and local on grasslands and sandy heaths in Denmark and southern Fennoscandia. Not recorded from Norway. – Holarctic, common in the Mediterranean countries.

Biology. Recorded from nests of Sphecidae: *Ammophila, Tachysphex*. Walton (1915) mentions two specimens bred from grasshoppers.


22. *Taxigramma elegantulum* (Zetterstedt, 1844)
    Figs 112, 113, 116, 123-127; pl. 1: 3.

*Tachina elegantula* Zetterstedt, 1844, Dipt. Scand., 3: 1024.

Male with silvery pollinosity and an abdominal pattern of 3 dorsal spots which almost coalesce. Female lighter, greyish pollinose; with well-defined abdominal spots.

♂. Head with frons protruding (Fig. 112). Frontal vitta, fronto-orbital plate, and parafacial plate silvery pollinose. Fronto-orbital plate with 2 proclinate orbital bristles, without additional hairs although a third hair-like orbital may be present above the others. Parafacial plate densely haired. Antennae blackish grey, second antennomere reddish apically. Third antennomere 2.5-3.0× as long as second. Arista without or with very short hairs, the longest not more than 0.3× greatest aristal diameter. Palpi brown to yellowish brown. Proboscis short. Thorax black, silvery grey pollinose, the area between the dorsocentrals slightly changing in colour with the incidence of light. Legs black, claws and pulvilli short. Tarsomeres 1-4 of fore tarsus with long sensory hairs on anteroventral surface. Last section of wing-vein $CuA_1$ as long as previous section (Fig. 116). Abdomen black with silvery grey pollinosity. T3-T5 each with 3 dorsal spots which coalesce into an irregular, transverse black spot. T3-T4 each with a pair of median marginal bristles, T5 with a row of marginals which has advanced dorsally to an almost discal position. Terminalia similar to *Hilarella stictica*.

♀. Somewhat like the male but generally lighter in colour. Frons less prominent, parafacial setulae shorter and much sparser (Fig. 113). Frontal vitta yellow, fronto-orbital and parafacial plates whitish grey. Antennae orange, with third antennomere blackish. Palpi yellow. Tarsi black, legs otherwise more or less yellowish. Fore tarsus with sensory hairs of normal length. Abdomen with 3 black spots on all tergites.

Length ♂♀. 4.5-6.0 mm.

Distribution. Common but local in dry sandy places in Denmark. In Fennoscandia north to Nb. in Sweden and ObS in Finland. Not recorded from Norway. – Widely distributed in Europe, east to Iran and Mongolia. Not in the British Isles.

Biology. Unknown. Kramer (1917) suggests that larval development takes place in nests of the ant *Formica cinerea* Mayr, but this needs confirmation.

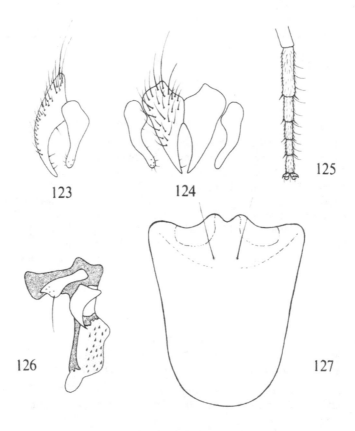

Figs 123-127. *Taxigramma elegantulum* (Zetterstedt). – 123: cerci + surstyli, lateral view; 124: cerci + surstyli, posterior view; 125: right fore tarsus ♂, dorsal view; 126: aedeagus; 127: ST5 ♂.

65

# Genus *Phrosinella* Robineau-Desvoidy, 1863

*Phrosinella* Robineau-Desvoidy, 1863, Hist. nat. Dipt. Paris, 2: 82 (as *Phrosina* on p. 101. Preocc., Risso, 1836).

Type species: *Phrosina argyrina* Robineau-Desvoidy, 1863, = *Tachina nasuta* Meigen, 1824.

Small to medium-sized flies. Head with profrons protruding somewhat like females of *Metopia*. Frontal vitta broad and only slightly narrowing towards lunula. Occiput with some hyaline hairs which may be restricted to the prestomal bridge or extended to postgena and gena. Fronto-orbital and parafacial plates with scattered setulae. Lunula setose. Palpi yellow.

Males of all Nearctic species and of the Palaearctic *P. sannio* possess distinctly modified fore tarsi with enlarged proximal segments and elongated hairs. Males of some species possess pictured wings and have a complex courtship behaviour (Spofford & Kurczewski 1985). Female fore tarsi are very broad and may be specialised for digging out closed entrances of the sphecid hosts (Allen 1926).

About 17 species, distributed in the Nearctic and Palaearctic regions.

## Key to species of *Phrosinella*

1    Head with some white hairs on prestomal bridge below occipital foramen. Genal and postgenal hairs black. Abdominal T3-T4 each with pollinosity extending to hind-margins and a pattern of 3 ill-defined black spots. Male fore tarsus strongly modified (Fig. 128) . . . . . . . . . . . . . . . . . . . . . . . . . . . . . 23. *sannio* (Zetterstedt)

–    Head with some white hairs on both prestomal bridge and postgena. Abdominal T3-T4 with pollinosity never reaching hind-margins and at most with a single median spot. Male fore tarsus unmodified . . . . . . . . . . . . . . . . . . . . . . . . . . . . 24. *nasuta* (Meigen)

## 23. *Phrosinella sannio* (Zetterstedt, 1838)
    Figs 128-132.

*Tachina sannio* Zetterstedt, 1838, Insecta Lapp.: 636.
*Tachina pilitarsis* Zetterstedt, 1844, Dipt. Scand., 3: 1021.
*Phrosinella septentrionalis* Rohdendorf, 1970, Opr. nasek. evrop. SSSR: 643.

Medium-sized species. Females brownish grey, males more blackish. Hyaline or white occipital hairs restricted to prestomal bridge. Abdomen with ill-defined spots. Male fore tarsus modified.

♂. Head: frons slightly protruding. Frontal vitta, fronto-orbital and parafacial plates silvery grey pollinose. Fronto-orbital plate haired, with 2 proclinate and 1 reclinate orbital bristles. Parafacial plate haired almost to lower margin of eye. Lunula setose.

First and second antennomeres reddish; third antennomere black, 3.5-4.0× as long as second. Arista thickened in proximal 0.6-0.7. Genal and postgenal hairs black except for a few hyaline or whitish hairs on prestomal bridge below the neck. Palpi yellow or brownish yellow, proboscis brown. Thorax black with sparse grey pollinosity. Legs black. Fore tarsus with stout first and second tarsomeres, the latter with a brush of setae on posterior surface. Tarsomeres 3-5 slender (Fig. 128). Abdomen black with grey pollinosity. T1 + 2 very sparsely pollinose, T3-T4 with 3 ill-defined black spots, T5 with pollinosity on anterior 0.4-0.5. T1 + 2-T3 with a pair of median marginal bristles,

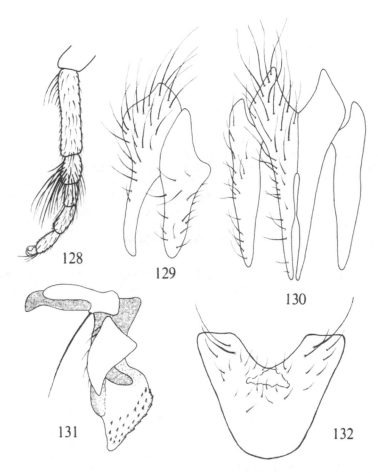

Figs 128-132. *Phrosinella sannio* (Zetterstedt). – 128: right fore tarsus ♂, posterior view; 129: cerci + surstyli, lateral view; 130: cerci + surstyli, posterior view; 131: aedeagus; 132: ST5 ♂.

T4-T5 with a row of marginals. Terminalia: aedeagus with divided dorsal plate and well sclerotised ventral plates (Fig. 131).

♀. Somewhat like the male but more brownish in colour. Frontal vitta olive pollinose. Tarsomeres of fore tarsus broadened. Abdominal pattern with a distinct median black spot on each of T1+2-T4. T5 with pollinosity on anterior 0.5.

Length ♂♀. 5.5-7.5 mm.

Distribution. Not found in Denmark. Common in sandy areas of Fennoscandia south of the Arctic Circle. – A boreal species: from Fennoscandia to Taymyr and the Far East.

Biology. Unknown.

24. *Phrosinella nasuta* (Meigen, 1824)
   Figs 133, 134.

*Tachina nasuta* Meigen, 1824, Syst. Beschr., 4: 374.

Small to medium-sized species with white hairs on postgenae. Abdominal T3-T5 each with a transverse band of silvery pollinosity.

♂. Head as in *P. sannio* but with white hairs covering all of occiput, posterior part of postgena, and often extending on to remaining postgena and gena. Frontal vitta olive pollinose. Thorax as in *P. sannio*. Male fore tarsus unmodified. Abdomen black, anterior part of T3-T5 each with a transverse band of silvery, often somewhat olive-tinged, pollinosity. Terminalia: cerci and surstyli like *P. sannio*. Aedeagus less swollen in lateral view but very broad in posterior view and with an unpaired sclerotisation ventrally (Figs 133, 134).

♀. Like the male but fore tarsus with broadened tarsomeres.
Length ♂♀. 4.5-6.5 mm.

Distribution. No records from Denmark; from Fennoscandia only a single female from Sweden: Vrm., S. Bengtsson leg. (ZML), and a female from USSR: Vib, Forsius leg. (ZMH). – Widely distributed in the Palaearctic region: from western Europe and North Africa to Mongolia. Not in the British Isles.

Biology. Unknown.

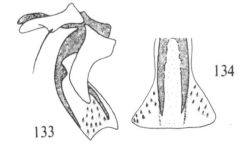

Figs 133, 134. *Phrosinella nasuta* (Meigen). – 133: aedeagus; 134: distiphallus, posterior view.

# Genus *Metopia* Meigen, 1803

*Metopia* Meigen, 1803, Mag. Insektenk., 2: 280.
  Type species: *Musca leucocephala* Rossi, 1790, preocc., = *Tachina argyrocephala* Meigen, 1824.

A very distinct genus of medium-sized flies, with moderately to strongly developed sexual dimorphism.

  Head profile almost triangular, proboscis short. Antennae long, lunula with setae, inner parafacial margin with a row of bristly setae close to facial ridge, and orbital

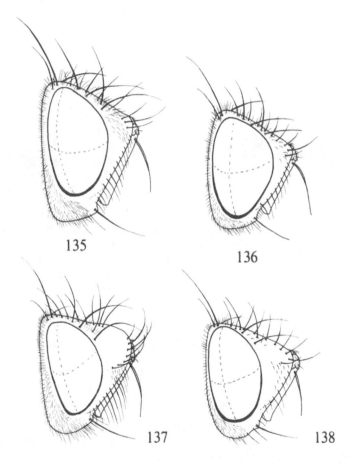

135

136

137

138

Figs 135-138. *Metopia* spp., head profile. – 135: *M. campestris* (Fallén) ♂; 136: *M. grandii* Venturi ♂; 137: *M. argyrocephala* (Meigen) ♂; 138: *M. argyrocephala* (Meigen) ♀.

Figs 139-141. *Metopia* spp. ♂, head in dorsal view. - 139: *M. argyrocephala* (Meigen); 140: *M. roserii* Rondani; 141: *M. tshernovae* Rohdendorf. (After Rohdendorf 1955.)

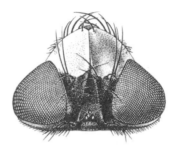

139

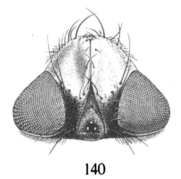

140                    141

bristles in an inner reclinate and an outer proclinate row. The last character is unique in the Miltogrammatinae, while the others occur sporadically within the subfamily.

Males are more hairy and may possess specialised tarsal setae. In the subgenus *Metopia* sensu stricto, the male profrons is very large and produced, with the anterior part of frontal vitta more or less obliterated by the contiguous fronto-orbital plates.

Note that several species of *Metopia* have only been recognized relatively recently (Venturi 1952; Rohdendorf 1955), which makes older records of *Metopia argyrocephala* rather unreliable. In addition, most females of *Metopia* sensu stricto are still impossible to separate on morphological grounds.

At present the genus contains about 36 species and is represented in all zoogeographical regions.

### Key to species of *Metopia*

| | |
|---|---|
| 1 | Mid tibia with 1 av bristle. Male fore tarsus with very long posterior hairs (Fig. 142) ........................ 25. *campestris* (Fallén) |
| – | Mid tibia without av bristles. Male fore tarsus different ................. 2 |
| 2(1) | Abdominal T1 + 2 without median marginal bristles .................... 3 |
| – | Abdominal T1 + 2 with a pair of median marginal bristles .............. 4 |

3(2) Ocellar bristles about as strong as reclinate orbitals. 2-4 su-
pra-vibrissal setae . . . . . . . . . . . . . . . . . . . . . . . . . . . . . . . 26. *grandii* Venturi
–     Ocellar bristles about as long as, but weaker than, reclinate
orbitals. 1-2 supra-vibrissal setae. . . . . . . . . . . . . . . . . . 30. *roserii* Rondani ♀
(Note that aberrant specimens of *M. argyrocephala* with the
median marginals of T1 + 2 lacking will key out here.)
4(2) Females . . . . . . . . . . . . . . . . . . . . . . . . . . . . . 27. *argyrocephala* (Meigen),
28. *staegerii* Rondani,
and 29. *tshernovae* Rohdendorf
–     Males . . . . . . . . . . . . . . . . . . . . . . . . . . . . . . . . . . . . . . . . . . . . . . . . . . . . . . 5
5(4) Fore tarsus with slightly elongated, somewhat curled hairs
posteriorly (Fig. 146) . . . . . . . . . . . . . . . . . . . . . . . . . . . 28. *staegerii* Rondani
–     Fore tarsus without elongated hairs posteriorly. . . . . . . . . . . . . . . . . . . . . . . . 6
6(5) Anterior part of fronto-orbital plates subcontiguous, frontal
vitta distinct to lunula (Fig. 141). Row of frontal bristles continuous, uninterrupted. Gonopods straight or only slightly
curved (Fig. 157). . . . . . . . . . . . . . . . . . . . . . . . . . 29. *tshernovae* Rohdendorf
–     Anterior part of fronto-orbital plates contiguous, totally
obliterating anterior part of frontal vitta (Figs 139, 140).
Row of frontal bristles interrupted or weakly developed
along the contiguous part of fronto-orbital plates . . . . . . . . . . . . . . . . . . . . . 7
7(6) Silvery part of fronto-orbital plate covering at least anterior
0.6, with a gradual transition to the posterior greyish part.
Row of frontal bristles almost uninterrupted, but weakly

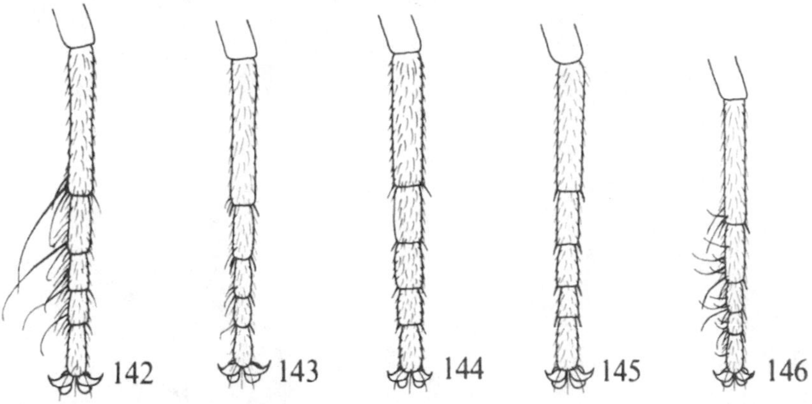

Figs 142-146. *Metopia* spp., right fore tarsus, dorsal view. – 142: *M. campestris* (Fallén) ♂; 143: *M. grandii* Venturi ♂; 144: *M. argyrocephala* (Meigen) ♀; 145: *M. argyrocephala* (Meigen) ♂; 146: *M. staegerii* Rondani ♂.

developed along contiguous part of fronto-orbital plates
(Fig. 140) ..................................... 30. *roserii* Rondani
– Silvery part of fronto-orbital plate covering anterior 0.4-0.5,
abruptly demarcated from the posterior greyish part (Fig.
139). Row of frontal bristles distinctly interrupted along
contiguous part of fronto-orbital plates ........ 27. *argyrocephala* (Meigen)

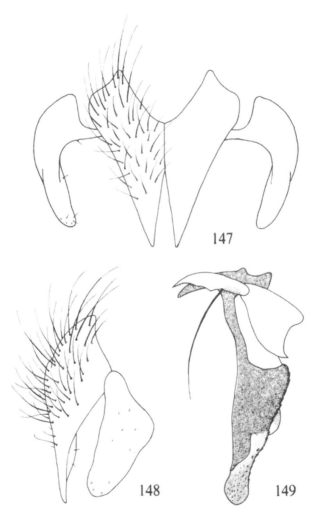

Figs 147-149. *Metopia campestris* (Fallén). – 147: cerci + surstyli, posterior view; 148: cerci + surstyli, lateral view; 149: aedeagus.

### 25. *Metopia campestris* (Fallén, 1810)
Figs 135, 142, 147-149.

*Tachina campestris* Fallén, 1810, K. svenska VetenskAkad. Handl., [2]31: 266.

A rather robust, bristly species with an almost triangular head-profile. Colour black with abdominal pattern of silvery pollinosity.

♂. Head almost triangular. Frontal vitta broad, at level of anterior ocellus about 3.3× width of one fronto-orbital plate. Frontal vitta black, only slightly grey to brownish grey pollinose. Lower 0.7 of fronto-orbital plates and all of parafacial plates silvery pollinose. Ocellar bristles about as strong as reclinate orbitals. Fronto-orbital plate with 2 outer proclinate and 2 inner reclinate orbitals, and numerous black hairs. Parafacial plate with a row of strong bristles on inner margin close to facial ridge, and on upper part with some black hairs which continue on to the fronto-orbital plate. Third antennomere 4.5-5.5× as long as second. Arista about as long as third antennomere. Vibrissae strong, with 2-4 supravibrissal setae. Thorax grey pollinose, legs black. Mid tibia with 1 ad, 1 av, and about 3 pd bristles. Very rarely with 2 ad bristles on mid tibia. Apical part of tarsomeres 1-4 of fore tarsus with long, forwardly-curved setae on posterior surface (Fig. 142). Abdomen with T1 + 2 black, T3-T4 silvery grey pollinose, each with 3 black triangular spots which coalesce at hind margin. T5 with silvery grey pollinosity at anterior 0.75. T1 + 2-T3 each with 1-2 pairs of median marginal bristles, T4-T5 with a row of marginals. Terminalia with cerci rather broad in posterior view, and gonopods short and slightly curved. Aedeagus with well-developed ventral plates and between these a median projection (Fig. 149).

♀. Like the male but with shorter and sparser hairing. Face broader, at lunula almost as broad as frons at vertex. Fore tarsus without specialised hairs, and abdominal T1 + 2 with only 1 pair of median marginal bristles.

Length ♂♀. 4.5-8.0 mm.

Distribution. Very common in Fennoscandia and Denmark except for the northern parts of Norway and Finland. – Holarctic and Oriental (India, Korea).

Biology. Recorded from nests of Pompilidae: *Pompilus;* Vespidae: *Arachnospila;* Sphecidae: *Ammophila, Gorytes, Larropsis, Sphex.*

In Denmark bred from nests of *Ammophila campestris* (Latreille), *Arachnospila trivialis* Dahlbom, and *Pompilus* sp. (Lundbeck 1927; Nielsen 1932).


### 26. *Metopia grandii* Venturi, 1953
Figs 136, 143, 150-152.

*Metopia grandii* Venturi, 1953, Boll. Ist. ent. Univ. Bologna, 19: 166.

Rather like the female of *M. campestris,* but well characterised by the lack of mid tibial av bristle, abdomen without median marginals on T1 + 2, and male fore tarsus with a few short irregular hairs posteriorly.

♂. Head almost triangular in profile. Face at lunula only slightly narrowed; frontal vitta broad, at level of anterior ocellus 3.5 × width of one fronto-orbital plate. Head chaetotaxy as in *M. campestris* but fronto-orbital plates with only few hairs in addition to the bristles. Mid tibia with 1 ad and a row of pd, but without av bristles. Apical part of tarsomeres 2-4 of fore tarsus with some elongated hairs on posterior surface (Fig. 143), but this character is somewhat variable. Abdomen with pollinosity as in *M. campestris*. T1 + 2 without median marginal bristles, T3 with 1 pair, and T4-T5 with a row of marginals. Terminalia with cerci slender in posterior view, gonopods broad with a small hook apically, and aedeagus with narrow ventral plates and a median projection between these (Fig. 152).

♀. Like the male but without specialised hairs on fore tarsus.

Length ♂♀. 4.5-6.0 mm.

Distribution. Somewhat rare in Denmark and Fennoscandia. In Finland north to ObN. Not recorded from Norway. – Palaearctic, from western Europe east to Japan. Not in the British Isles.

Biology. Unknown.

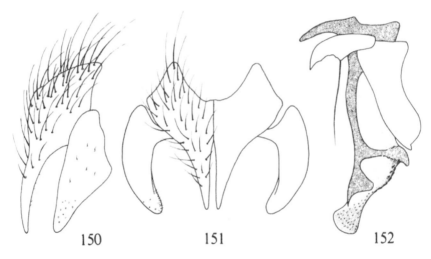

150          151          152

Figs 150-152. *Metopia grandii* Venturi. – 150: cerci + surstyli, lateral view; 151: cerci + surstyli, posterior view; 152: aedeagus.

### 27. *Metopia argyrocephala* (Meigen, 1824)
Figs 137-139, 144, 145, 153-156; pl. 1:4.

*Musca leucocephala* Rossi, 1790, Fauna etrusca, 2: 306. Preocc. by Villers, 1789.
*Tachina argyrocephala* Meigen, 1824, Syst. Beschr., 4: 372.

Strongly sexually dimorphic species. Males with profrons strongly produced, giving the head an almost trapezoidal profile. Female head more triangular in profile. Male abdominal pattern with an olive-brown tinge along the median spots, and lateral spots changing in colour with the incidence of light.

♂. Profrons conically protruding (Fig. 137), with a silvery pollinosity which is abruptly demarcated from the greyish pollinose posterior 0.5-0.6 of fronto-orbital plate. A break in reflectance occurs just below level of aristal insertion. Frontal vitta black, narrowing towards profrons where the fronto-orbital plates are contiguous, obliterating anterior 0.3 of frontal vitta. Ocellars weaker than reclinate orbitals. Two outer proclinate orbitals and 2 inner reclinate orbitals. Row of frontals absent along contiguous margins of fronto-orbital plates, but 1-2 bristles occasionally present. Parafacial plate with the usual row of bristles along inner margin close to facial ridge, and with some black hairs on upper part. Third antennomere 5-6× as long as second. Vibrissae strong, with 1-2 supravibrissal setae. Thorax grey pollinose, legs black. Mid tibia with a strong ad and an irregular row of pd, but without the av bristle. Fore tarsus without specialised hairs. Abdomen with T1+2 black and almost without pollinosity. T3-T5 silvery pollinose, in posterior view with 3 black spots which coalesce at posterior margin; the median spot is triangular with brownish margins, the lateral spots almost rectangular. Abdominal pattern in dorsal/anterior view black medially with silvery spots laterally. T1+2 each with a pair of median marginal bristles, T5 with a row of marginals. Terminalia: aedeagus with narrow ventral plates and sickle-shaped gonopods (Fig. 155).

♀. Profrons conical, but much less so than in male, and without the distinct silvery spot (Fig. 138). Frontal vitta distinct to lunula, not obliterated by the fronto-orbital plates. Fore tarsus with sligthly broader tarsomeres (Fig. 144). Abdomen less changeable in colour with the incidence of light. The marginal bristles on T1+2 occasionally rather weak or totally absent. In the latter case, the specimens are indistinguishable from females of *M. roserii.*

Length ♂♀. 6.0-7.5 mm.

Distribution. Common in Denmark and Fennoscandia, reaching T.Lpm. in Sweden. – Widely distributed in the Palaearctic, Nearctic, northern Neotropical, and Oriental regions.

Biology. Recorded from nests of Vespidae: *Stenodynerus;* Sphecidae: *Ammophila, Bembix, Cerceris, Chlorion, Crossocerus, Encopognathus, Mellinus, Oxybelus, Philantus, Spheclus, Sphex;* Apidae: *Halictus, Lasioglossum.*

In Finland, Fahlander (1954) observed *M. argyrocephala* following females of *Argogorytes fargei* (Schuckard) when these returned with prey.

Note. As some infraspecific variation exists in the extent of the silvery part of profrons and in the presence or absence of frontal bristles along the contiguous part of the fronto-orbital plates, this species may sometimes be confused with *M. roserii.* Similarly, newly-hatched males may have the fronto-orbital plates slightly separated and may be taken for a *M. tshernovae.* In this case, the shape of the gonopods may be a reliable character for separating these species, as given in the key.

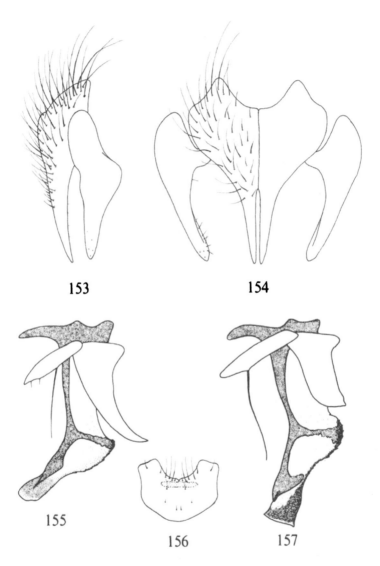

153

154

155

156

157

Figs 153-156. *Metopia argyrocephala* (Meigen). – 153: cerci + surstyli, lateral view; 154: cerci + surstyli, posterior view; 155: aedeagus; 156: ST5 ♂.
Fig. 157. *Metopia tshernovae* Rohdendorf, aedeagus.

28. *Metopia staegerii* Rondani, 1859
   Fig. 146.

*Metopia staegerii* Rondani, 1859, Dipt. Ital. Prodromus, 3: 210.
*Metopia rondaniana* Venturi, 1953, Boll. Ist. ent. Univ. Bologna, 19: 163.

Very like *M. argyrocephala* in general appearance, but easily distinguished in the male sex by the specialised hairs of fore tarsus: the apical part of tarsomere 1 and all of tarsomeres 2-4 have somewhat curled hairs posteriorly and posterodorsally (Fig. 146). Terminalia similar to those of *M. argyrocephala,* with sickle-shaped gonopods.
   ♀. At present indistinguishable from other females of *Metopia* s. str.
   Length ♂♀. 6.0-7.5 mm.

Distribution. Common in Denmark and southern parts of Sweden and Finland. Only a few records from southern Norway. – Western Palaearctic east to the Altai Mountains. Not in the British Isles.

Biology: Unknown.


29. *Metopia tshernovae* Rohdendorf, 1955
   Figs 141, 157.

*Metopia tshernovae* Rohdendorf, 1955, Ént. Obozr., 34: 368.

Very similar to *M. argyrocephala,* but anterior part of fronto-orbital plates subcontiguous, leaving a rather narrow space for the frontal vitta, and the row of frontals almost complete and symmetrical. What appear to be intermediate specimens, or specimens of *M. argyrocephala* in which the fronto-orbital plates have failed to meet completely, may occur. Such specimens should be identified by reference to the gonopods, which are straight (or only slightly curved) and broad in *M. tshernovae* (Fig. 157), distinctly curved in *M. argyrocephala* (Fig. 155).
   ♀. At present indistinguishable from other species of *Metopia* s. str.
   Length ♂♀. 6.0-7.5 mm.

Distribution. Rare but widely distributed in Denmark. Recorded from southeastern Norway and from Ab in Finland. Not recorded from Sweden. – Palaearctic and Oriental (Thailand), rather rare. Not in the British Isles.

Biology. Unknown.


30. *Metopia roserii* Rondani, 1859
   Fig. 140.

*Metopia roserii* Rondani, 1859, Dipt. Ital. Prodromus, 3: 210.
*Metopia instruens* Walker, 1859, J. Proc. Linn. Soc. Lond. Zool., 4: 129.
*Metopia stackelbergi* Rohdendorf, 1955, Ént. Obozr., 34: 369.

Similar to *M. argyrocephala* in general appearance. The male of *M. roserii* can be distinguished by the greater extent of the silvery pollinosity on profrons, the presence of some frontal bristles along the contiguous margins of fronto-orbital plates, and the weakly developed median abdominal spots. These characters, however, are somewhat variable, but in typical specimens the silvery part of frons covers at least anterior 0.6 of fronto-orbital plates, and the transition to the posterior greyish part is perfectly gradual (Fig. 140). One or 2 pairs of frontals are situated along the contiguous margins of the fronto-orbital plates, the number often being uneven. The median abdominal spot of T3-T4 is narrow, but still with a distinct brown tinge. Terminalia similar to those of *M. argyrocephala*.

♀. As the female of *M. argyrocephala* but without median marginal bristles on abdominal T1 + 2. This will not, however, ensure a definite identification, as female specimens of *M. argyrocephala* without marginals on T1 + 2 occur.

Length ♂♀. 6.0-7.5 mm.

Distribution. Recorded from southern Finland. No records from the rest of Fennoscandia or Denmark. – Palaearctic, east to Japan. Not in the British Isles. Oriental.

Biology. Bred from nests of Pompilidae: *Batozonus* (Verves 1981b).

## Genus *Macronychia* Rondani, 1859

*Macronichia* Rondani, 1859, Dipt. Ital. Prodromus, 3: 229 (222).
*Macronychia* Rondani, 1859, Dipt. Ital. Prodromus, 3: 239.
   Type species: *Tachina agrestis* Fallén, 1810, sensu Rondani, 1859; misidentification, = *Xysta striginervis* Zetterstedt, 1838.

Medium-sized to large grey species. Frons somewhat protruding, parafacial plate broad and setose. Vibrissae well-developed and situated high above lower head margin. Genal groove well-developed. Antennae short, third antennomere 1-2× as long as second. Arista pubescent or with short hairs. Mid tibia with 2 ad bristles. Claws and pulvilli of all legs longer than fifth tarsomere. Metathoracic spiracle with small posterior lappet.

The genus, which contains about 13 species, is widely distributed, but has not yet been recorded from the Australian region. The species are kleptoparasites, mainly of stem- and stalk-nesting Sphecidae, but the record of *M. polyodon* from nests of *Bombus* sp. (Verves 1982b), and of *Macronychia* sp. bred from adult tabanids (Thompson 1978), may indicate a broader host-range.

Note. The different spellings of the generic name in the original publication have given rise to some difference of opinion as to the correct version. Rohdendorf (1967) and Hackman (1980) use "*Macronichia*" while most other authors use the latinised form "*Macronychia*". I regard Verves (1982b, 1983) as the first revisor, and I follow his use of the latter spelling.

1  Basicosta yellow. Abdomen almost entirely grey or olive-
   grey, with a narrow median dark stripe (Fig. 167) ...... 32. *griseola* (Fallén)
-  Basicosta brown to black. All abdominal tergites with 3 black
   spots or stripes ................................................. 2
2(1)  ♂: cerci in posterior view with the prongs very close together
   (Fig. 172). ♀: terminalia modified into a distinct ovipositor
   (Fig. 174) ................................. 34. *striginervis* (Zetterstedt)
-  ♂: cerci in posterior view with widely separated prongs

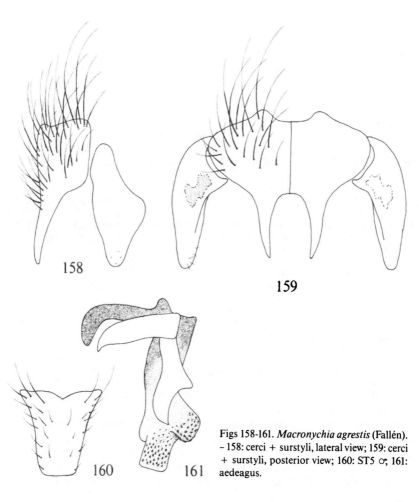

158

159

160   161

Figs 158-161. *Macronychia agrestis* (Fallén).
– 158: cerci + surstyli, lateral view; 159: cerci
+ surstyli, posterior view; 160: ST5 ♂; 161:
aedeagus.

79

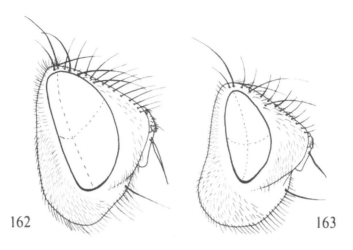

Figs 162, 163. *Macronychia* spp., head profile ♂. - 162: *M. agrestis* (Fallén); 163: *M. alpestris* Rondani.

(Fig. 158). ♀ : terminalia unmodified . . . . . . . . . . . . . . . . . . . . . . . . . . . . . . . . 3
3(2)  Parafacial plate very broad, 0.37-0.44 × eye-height, and with
4-6 irregular rows of setae (Fig. 163) . . . . . . . . . . . . . . . . . [*alpestris* Rondani]

*Macronychia alpestris* Rondani, 1865 [ = *Macronychia dumosa* (Pandellé, 1895)] is recorded from Finland by Hackman (1980) (as *M. conica* Robineau-Desvoidy), based on a single specimen in ZMH with no locality label (Pape in press). *Macronychia alpestris* is known from southern and central Europe, north to East Germany, Poland, and Lithuania, and east to Central Asia.

–  Parafacial plate narrower, 0.22-0.37 × eye-height, and with
1-3 irregular rows of setae (Fig. 162) . . . . . . . . . . . . . . . . . . . . . . . . . . . . . . . . 4
4(3)  Abdominal T1 + 2 with median marginal bristles . . . . . . . 31. *agrestis* (Fallén)
–  Abdominal T1 + 2 without median marginal bristles . . 33. *polyodon* (Meigen)

### 31. *Macronychia agrestis* (Fallén, 1810)
Figs 158-161, 162, 166.

*Tachina agrestis* Fallén, 1810, K. VetenskAkad. Handl., [2]31: 270.

Grey species with 3 black stripes on thorax and 3 rectangular to triangular black spots on abdominal tergites. Basicosta brown.
♂. Head: fronto-orbital and parafacial plates grey pollinose, frontal vitta black. Genal groove dark reddish. Fronto-orbital plate haired, with 2 proclinate and 1 reclinate orbital bristles. Parafacial plate densely haired. Antennae black. Third antennomere 1.5-2.0 × as long as second. Arista micropubescent, thickened in proximal 0.3, and 1.5 × as long as third antennomere. Palpi black. Proboscis of medium length.

Thorax grey pollinose with 3 black stripes dorsally. Legs black with thin grey pollinosity. Wings hyaline, basicosta brown. Abdomen grey pollinose with 3 rectangular to slightly triangular black spots on each of T1 + 2-T5. T1 + 2 with 1-2 pairs of median marginal bristles, T3-T5 each with a row of marginals. Terminalia with cerci broadly separated (Fig. 158). Aedeagus with a well-developed epiphallus and almost straight parameres.

♀. Like the male. Without an ovipositor.

Length ♂♀. 7.0-11.5 mm.

Distribution. Not rare in Denmark and southern parts of Fennoscandia. – Western Europe, except for the British Isles. East to Altai Mountains.

Biology. Recorded from nests of Sphecidae: *Psenulus*. A record of *M. agrestis* as a parasitoid in *Macrothylacia rubi* (Linnaeus) (Lepidoptera: Lasiocampidae) needs confirmation (Saager 1959).

### 32. *Macronychia griseola* (Fallén, 1820)
Figs 167, 168-171.

*Tachina griseola* Fallén, 1820, Monogr. Musc. Sveciae: 10.

Grey species. Basicosta yellow. Abdomen almost unicolorous grey, but with a narrow, median olive-brown stripe.

♂. Head: fronto-orbital and parafacial plates grey pollinose, frontal vitta black. Genal groove dark reddish. Fronto-orbital plate with 2 proclinate and 1 reclinate orbital bristles, with a few hairs at vertex, and densely haired anteriorly. Parafacial plate densely haired. Antennae black, second antennomere somewhat reddish on apical margin. Third antennomere as long as second. Arista with very short hairs, thickened

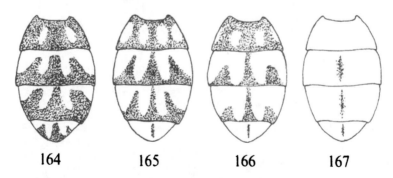

164          165          166          167

Figs 164-167. *Macronychia* spp., semidiagrammatical drawings of abdomen, dorsal view. – 164: *M. polyodon* (Meigen); 165: *M. striginervis* (Zetterstedt); 166: *M. agrestis* (Fallén); 167: *M. griseola* (Fallén).

in proximal 0.4, and 1.8 × as long as third antennomere. Palpi black. Proboscis short. Thorax densely pollinose, postsutural area with a narrow median olive-brown stripe. Abdomen almost unicolorous grey, with a narrow median olive-brown stripe. T3 with a pair of median marginal bristles. T4-T5 each with a row of marginals. Terminalia similar to those of *M. agrestis*.

♀. Like the male, but abdomen slightly more olive-grey. Without an ovipositor. Length ♂♀. 4.0-7.5 mm.

Distribution. Somewhat rare in Denmark, Sweden, and Finland. Not recorded from Norway. – Widely distributed in the Palaearctic and Oriental regions, from the British Isles east to China and Taiwan.

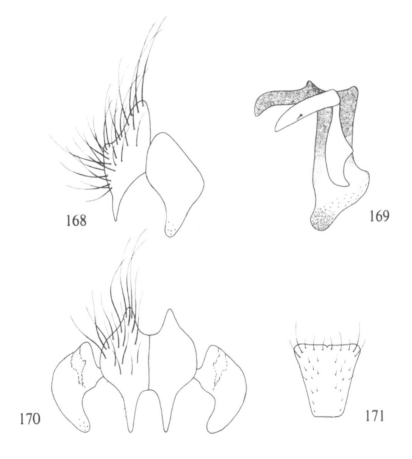

Figs 168-171. *Macronychia griseola* (Fallén). - 168: cerci + surstyli, lateral view; 169: aedeagus; 170: cerci + surstyli, posterior view; 171: ST5 ♂.

Biology. Recorded from nests of Sphecidae: *Oxybelus.*

## 33. *Macronychia polyodon* (Meigen, 1824)
Fig. 164.

*Tachina polyodon* Meigen, 1824, Syst. Beschr., 4: 302.

Grey species with 3 black to olive-brown stripes on thorax, and 3 triangular brown spots on abdominal tergites. Basicosta dark brown.

♂. Head as in *M. agrestis,* thorax slightly more olive-pollinose. Wings often slightly fumose. Abdomen with 3 brown or black triangular spots on all tergites, the spots coalescing at hind-margin. T1 + 2 without distinct median marginal bristles, T3 with or without a pair of median marginals, T4-T5 each with a row of marginals. Terminalia like those of *M. agrestis.*

♀. Like the male. Without an ovipositor.

Length ♂♀. 5.5-11.0 mm.

Distribution. Not rare in South Sweden. Only few records from Norway: VA, and Finland: Ab. Recorded from Denmark in Verves (1982b). – Palaearctic, from the British Isles east to Japan.

Biology. Recorded from nests of Sphecidae: *Crabro, Crossocerus, Ectemnius, Oxybelus, Pemphredon;* Apidae: *Bombus.*

## 34. *Macronychia striginervis* (Zetterstedt, 1838)
Figs 165, 172-174; pl. 2:1.

*Xysta striginervis* Zetterstedt, 1838, Insecta Lapp.: 633.

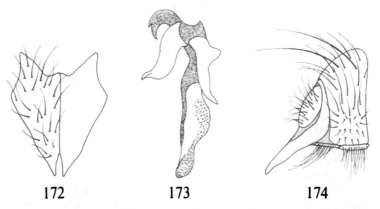

172          173          174

Figs 172-174. *Macronychia striginervis* (Zetterstedt). – 172: cerci in posterior view; 173: aedeagus; 174: terminalia ♀, lateral view.

83

Similar to *M. agrestis,* but easily separated by the following differences: wings often fumose along crossvein dm-cu and the bent part of vein M. Abdominal spots more triangular and just coalescing at posterior margin. Male cerci close together, parameres more or less S-shaped, and female terminalia modified into an ovipositor.

Length ♂♀. 7.0-13.0 mm.

Distribution. Common in eastern Denmark and southern parts of Fennoscandia. - Widely distributed in the Palaearctic and Afrotropical regions.

Biology. Recorded from nests of Sphecidae: *Ectemnius.* A single male in ZMUC was bred from a nest of *E. cavifrons* (Thomson).

## SUBFAMILY PARAMACRONYCHIINAE

A group of medium-sized to large species. Female frons broad and with proclinate orbital bristles, male frons narrow and without proclinate orbitals, except in *Sarcophila* where the frons is equibroad in both sexes. Antennal arista bare or haired. Notopleuron with 2 strong bristles, and often with additional hairs. Katepimeron separated from meron (coxopleural streak present). Mid tibia with 2 strong ad bristles. Hind coxa bare on posterior surface. Male terminalia with 2 segments. Protandrial segment, resulting from the fusion of T6 and syntergosternite 7 + 8, distinctly angular in profile (Fig. 14); a row of strong bristles marks the segmental origin of the sclerite.

The biology is diverse, varying from necrophagy and predation of snails to parasites/parasitoids of mammals and various insects.

Note that *Eurychaeta* Brauer & Bergenstamm (*Helicobosca* Bezzi) has recently been transferred from the Sarcophagidae to the Calliphoridae (Rognes 1986b).

## Genus *Agria* Robineau-Desvoidy, 1830

*Agria* Robineau-Desvoidy, 1830, Essai Myod.: 376.
   Type species: *Agria punctata* Robineau-Desvoidy, 1830.
*Pseudosarcophaga* Kramer, 1908. Ent. Wbl., 25: 200.
   Type species: *Musca affinis* Fallén, 1817, preocc., = *Agria punctata* Robineau-Desvoidy, 1830.

Dark, greyish pollinose species. Male with narrow frons, 0.06-0.15 × head-width, and without proclinate orbital bristles. Females with broader frons, about 0.3 × head-width, and with 2 proclinate orbitals. Parafacial plate narrow, with a row of setae along outer margin near eye. Arista haired. Occipital and postgenal hairs black. Abdomen with indistinct lateral spots and a median stripe.

*Agria* is a Holarctic genus with 1 Nearctic and 3 Palaearctic species. The larvae are predators of butterfly and moth pupae, occasionally attacking sawfly pupae.

*Agria punctata* and *A. mamillata* are very similar, especially in the female sex, and misidentifications may be expected in the older literature. In addition, females of *Ag-*

*ria* may be taken for females of Sarcophaginae and will key out to *Pierretia sexpunctata* (Fabricius) in the key. *Agria* is easily separated from the latter by the reduced costal spine, the short parafacial setae, the lack of hairs on posterior part of hind coxa, and the structure of the terminalia (T6 undivided).

## Key to species of *Agria*

1  ♂: protandrial segment and epandrium without paired knobs.
   Aedeagus as in Fig. 177. ♀: epiproct with numerous (14-18)
   setae (Fig. 179); never divided . . . . . . . . . . . . . . 35. *punctata* Robineau-Desvoidy
-  ♂: protandrial segment and epandrium with paired knobs
   (Fig. 187). Aedeagus as in Fig. 185. ♀: epiproct often divided,
   with 4 long setae and 0-4 (-8) additional short hairs (Fig. 181)
   . . . . . . . . . . . . . . . . . . . . . . . . . . . . . . . . . . . . . . 36. *mamillata* (Pandellé)

35. *Agria punctata* Robineau-Desvoidy, 1830
    Figs 175-179.

*Musca affinis* Fallén, 1817, K. VetenskAkad. Handl., (3) 1816: 237. Preocc. by Turton, 1800 and Lamarck, 1816.
*Agria punctata* Robineau-Desvoidy, 1830, Essai Myod.: 377.

Very similar to *A. mamillata* from which it can only be distinguished by the terminalia as described in the key.

♂. Fronto-orbital and parafacial plate silvery grey. Frontal vitta black, broadest at lunula. Narrowest part of frons 0.12 × head-width. No orbital bristles; inner verticals well-developed, outer verticals as long as adjacent postocular setae. Ocellar bristles rather weak. Parafacial plate with a somewhat irregular row of setae along outer margin near eye. Genal groove large. Occiput and postgena covered with black hairs. Antennae black, third antennomere 2 × as long as second. Arista with long hairs in proximal 0.6, second aristomere as long as broad. Palpi and proboscis black. Thorax grey pollinose. Presutural part with 5 black stripes, 3 of which continue on to postsutural part. Legs black, mid femur with an apical pv comb. Claws and pulvilli longer than fifth tarsomere. Wings hyaline, basicosta yellow. Costal spine indistinct. Lower calypters brownish in centre. Abdomen grey pollinose. T1+2 almost black, T3-T5 each with a median black stripe and a pair of lateral black spots. T3-T4 each with a pair of median marginal bristles, T5 with a row of marginals. ST5 rather small (Fig. 178). Terminalia protruding, grey pollinose. Protandrial segment and epandrium without paired knobs.

♀. Frons broad, at vertex 0.30-0.33 × head-width, broadening towards lunula. Frontal vitta parallel or slightly widening towards lunula. Two proclinate orbitals, the anterior one strongest. Outer verticals well-developed, ocellars strong. Terminalia differing from those of *A. mamillata* in the structure of the epiproct, which always has numerous (14-18) setae (Fig. 179), but see discussion under *A. mamillata*.
Length ♂♀. 6.0-9.0 mm.

Distribution. Common in Denmark and Sweden north to Vrm. Recorded from southern Finland. No records from Norway. – Western Europe and the British Isles, east to Mongolia.

Biology. The larvae are predators of lepidopterous pupae: *Aphelia, Aporia, Arctia, Autographa, Cacoecia, Dendrolimus, Euproctis, Hyphantria, Leucoma, Lymantria, Malacosoma, Vanessa;* and of sawfly pupae: *Empria, Diprion.* Records from beetles and grasshoppers (Séguy 1932) need confirmation.

Many records of *A. punctata* bred from ermine moths (*Yponomeuta* spp.) refer to *A. mamillata,* and this is probably also true of the record of Forsius (1924) who found larvae of *Agria* associated with *Yponomeuta evonymella* (Linnaeus) in Finland.

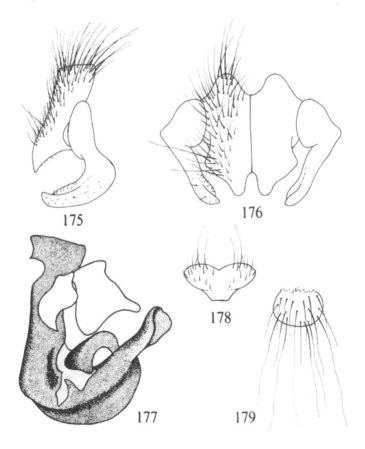

Figs 175-179. *Agria punctata* Robineau-Desvoidy. – 175: cerci + surstyli, lateral view; 176: cerci + surstyli, posterior view; 177: aedeagus; 178: ST5 ♂; 179: epiproct ♀.

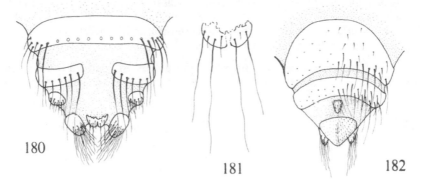

Figs 180-182. *Agria mamillata* (Pandellé), terminalia ♀. – 180: dorsal view; 181: epiproct; 182: ventral view.

Reported from Denmark as a predator of *Leucoma salicis* (Linnaeus) (Lymantriidae) (Nielsen 1914) and *Anticlea derivata* (Denis & Schiffermüller) (Geometridae) (Lundbeck 1927). There is a female from Sweden, Sk. (ZML) bred from a gipsy moth (*Lymantria* sp.). Bred from *Autographa gamma* (Linnaeus) (Noctuidae) in Finland.

The record in Zetterstedt (1859) of *A. punctata* bred from a dead cockchafer, *Melolontha melolontha* (Linnaeus), needs confirmation.

### 36. *Agria mamillata* (Pandellé, 1896)
Figs 180-187.

*Sarcophila mamillata* Pandellé, 1896, Revue ent., 15: 172.

Very similar to *A. punctata* and only distinguishable by the structure of the terminalia.

Males are easily recognised by the paired knobs on both protandrial segment and epandrium (Fig. 187), which are visible without dissection, and by the aedeagal structure (Fig. 185). Females may be recognised by the sparsely haired epiproct, which has 4 long setae and 0-8 short hairs (Fig. 181). The epiproct of *A. punctata* has about 14-18 setae (Fig. 179).

The female terminalia of *A. punctata* and *A. mamillata* seem to be somewhat variable. Verves (1982b) separates the species on the following characters:

*A. punctata:* the two sclerites of T7 each with 3-4 marginal setae; T8 missing; epiproct (T10) semicircular, setose, and indented apically.

*A. mamillata:* the two sclerites of T7 each with 6-8 marginal setae; T8 present as two weak sclerites; epiproct divided into two oval sclerites, each with a pair of setae.

I have been unable to confirm the differences in T7-T8. The number of marginal setae on T7 seems to be rather variable, and both species have T8 divided into two weak sclerites with a few setae. The division of the epiproct in *A. mamillata* also seems to be

87

of limited value in species recognition as specimens of *A. mamillata* occur with undivided epiproct.

The number of setae on the epiproct is rather constant and at present seems to be the best character for the definitive separation of the species.

Distribution. In Denmark only from NEZ. Recorded from southern Norway and southern Finland, in Sweden north to Vrm. – Palaearctic region, from western Europe to the Far East.

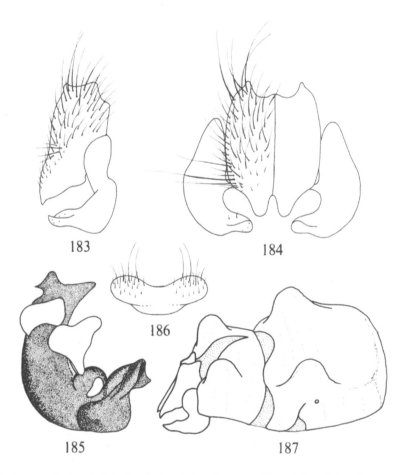

Figs 183-187. *Agria mamillata* (Pandellé). – 183: cerci + surstyli, lateral view; 184: cerci + surstyli, posterior view; 185: aedeagus; 186: ST5 ♂; 187: terminalia ♂ showing tubercles on protandrial segment and epandrium.

Biology. *Agria mamillata* seems to be restricted to final-instar larvae and pupae of ermine moths (*Yponomeuta* spp.). Each fly-larvae consumes about 3-4 pupae before hibernating (Nielsen 1914; Junnikkala 1960). It is a predator of *Y. evonymella* (Linnaeus) and *Y. malinellus* Zeller in Finland (Tiensuu 1939; Junnikkala 1960; Pyörnila & Pyörnila 1979). There is a Danish specimen in ZMUC bred from *Y. padella* (Linnaeus). Zetterstedt's (1845) record of *Agria punctata* bred from *Y. padella* in Sweden probably refers to *A. mamillata*.

## Genus *Sarcophila* Rondani, 1856

*Sarcophila* Rondani, 1856, Dipt. Ital. Prodromus, 1: 86.
   Type species: *Musca latifrons* Fallén, 1817.

Grey or olive-grey species with weakly developed thoracic stripes. Arista haired, parafacial plate with 1-3 rows of hairs. Both sexes with proclinate orbital bristles and

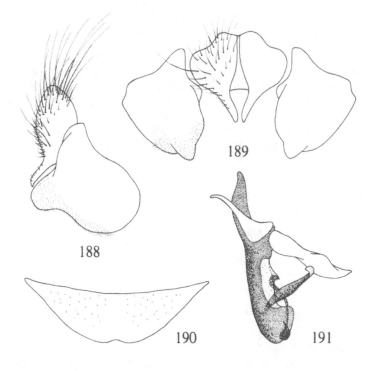

Figs 188-191. *Sarcophila latifrons* (Fallén). – 188: cerci + surstyli, lateral view; 189: cerci + surstyli, posterior view; 190: ST5 ♂; 191: aedeagus.

broad frons. Abdomen with more or less distinct black spots. Male terminalia distinctly protruding.

The genus contains several Palaearctic species, but many are only recently described (Verves 1982b, 1985), and distribution patterns are poorly known. The larvae are necrophagous or are insect parasitoids.

### 37. *Sarcophila latifrons* (Fallén, 1817)
Figs 188-191.

*Musca latifrons* Fallén, 1817, K. VetenskAkad. Handl., [3]1816: 238.

Densely grey pollinose with 3 black triangular spots on abdominal tergites. Both sexes with broad frons and 2 pairs of proclinate orbital bristles.

♂. Fronto-orbital plate grey. Frontal vitta black in posterior view, light grey in anterior view. Parafacial plate light grey. Frons at vertex 0.40-0.50 × head-width. Frontal vitta broad, parallel, at level of anterior ocellus 2.0-2.3 × width of one fronto-orbital plate. Two proclinate orbital bristles and a few scattered fronto-orbital hairs. Parafacial plate with hairs arranged in 2 more or less well-defined rows. Antennae black except for the reddish apical margin of second antennomere. Third antennomere 1.6-2.0 × length of second. Arista with long hairs in proximal 0.4-0.6. Palpi and proboscis black. Thorax grey, legs black with sparse grey pollinosity. Legs bristly, mid tibia with 4 ad bristles increasing in length distally, 2-4 pd, 2 p, and a strong v. Claws and pulvilli about 0.75 × as long as fifth tarsomere. Abdomen grey or slightly olive-grey, each tergite with 3 dorsal triangular black spots which may, however, be absent on T5. T3 with a weak or indistinct pair of median marginal bristles, T4-T5 with a row of marginals. Terminalia protruding, grey or olive pollinose.

♀. Like the male, but with frontal vitta at level of anterior ocellus 2.7-3.0 × width of one fronto-orbital plate.

Length ♂♀. 4.5-8.5 mm.

Distribution. Common in Denmark and southern Sweden. Listed from Norway and Finland in Verves (1986). – Palaearctic, from British Isles to East Siberia.

Biology. Up-to-date and reliable information is badly needed. Often cited as breeding in animal carcasses and dead insects (e.g. Sajo 1898; Séguy 1941; Emden 1954). Recorded from egg-pods and adults af acridid grasshoppers (Séguy 1932), but probably misidentified.

## Genus *Angiometopa* Brauer & Bergenstamm, 1889

*Angiometopa* Brauer & Bergenstamm, 1889, Denkschr. Akad. Wiss. Wien, Kl. math.-naturw., 56(1): 123.
Type species: *Musca ruralis* Fallén, 1817, preocc., = *Angiometopa falleni* Pape, 1986.

Medium-sized to large species with haired arista. Parafacial plate with some hairs.

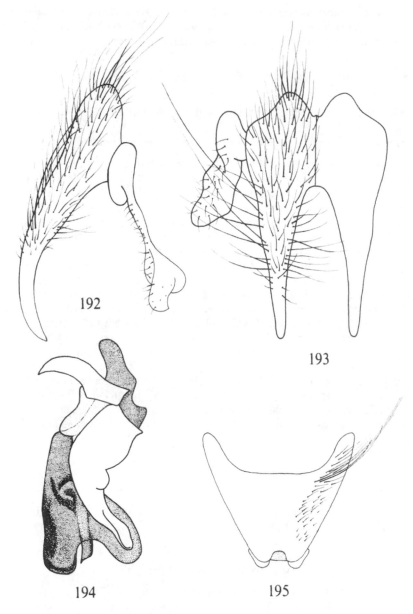

Figs 192-195. *Angiometopa falleni* Pape. – 192: cerci + surstyli, lateral view; 193: cerci + surstyli, posterior view; 194: aedeagus; 195: ST5 ♂.

Males without proclinate orbital bristles, frons narrow. Females with 2 proclinate orbitals and broader frons. Abdominal tergites each with 3 distinct black spots.

The genus is represented in the Palaearctic region by 5 species.

### 38. *Angiometopa falleni* Pape, 1986
Figs 192-195.

*Musca ruralis* Fallén, 1817, K. VetenskAkad. Handl., [3]1816: 236. Preocc. by Gravenhorst, 1807.
*Angiometopa falleni* Pape, 1986: 306. New name for *Musca ruralis* Fallén, 1817.

Medium-sized to large grey species. Males without orbital bristles, females with 2 proclinate orbitals. Palpi yellow. Abdominal tergites with 3 distinct black spots.

♂. Fronto-orbital and parafacial plates light grey pollinose; frontal vitta black, in anterior view with sparse grey pollinosity. Frons at narrowest point 0.19-0.23 × head-width. Frontal vitta at level of anterior ocellus about 2 × width of one fronto-orbital plate. No orbital bristles. Fronto-orbital plate sparsely haired, parafacial plate with some hairs in 2 ill-defined rows. First and second antennomeres reddish; third antennomere black or slightly reddish basally, 1.5-2.0 × as long as second. Arista haired in proximal 0.75. Palpi reddish to yellow. Thorax grey pollinose, with 3 black stripes, and with distinct prst acr. Legs black, grey pollinose. Mid femur with an apical pv comb. Claws and pulvilli longer than fifth tarsomere. Basicosta yellow, lower calypters infuscated in centre. Abdomen densely grey pollinose; tergites each with 3 more or less triangular black spots. The median spot of T5 absent or rather weak. T3 with a pair of weak median marginal bristles, T4-T5 with a row of marginals. Terminalia grey pollinose. Ground-colour varying from black to reddish.

♀. Frons broad, at vertex 0.33-0.40 × head-width. Frontal vitta about 2 × width of one fronto-orbital plate. Two pairs of proclinate orbitals. Mid femur without an apical pv comb, claws and pulvilli slightly shorter than fifth tarsomere. Abdomen with somewhat smaller lateral spots. Terminalia reddish. T7 with a row of strong marginal bristles.

Length ♂♀. 6.0-9.0 mm.

Distribution. Not rare in Denmark and South Sweden. No records from Norway. In Finland from the southern and eastern provinces. – Palaearctic, east to Mongolia.

Biology. Bred from pupae of *Lymantria monacha* (Linnaeus) (Lepidoptera: Lymantriidae). Larvae have also been found in wounds in horses and humans (Séguy 1941).

## Genus *Brachicoma* Rondani, 1856

*Brachicoma* Rondani, 1856, Dipt. Ital. Prodromus, 1: 69.
Type species: *Tachina nitidula* Meigen, 1824, sensu Rondani, 1856; misidentification, = *Tachina devia* Fallén, 1820.
*Brachycoma;* erroneous subsequent spelling.

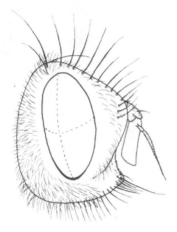

Fig. 196: *Brachicoma devia* (Fallén), head profile ♂.

Medium-sized to large blackish species. Frons and lower facial margin somewhat protruding. Females with broad frons and 2 proclinate orbital bristles, males with narrower frons and without orbitals. Parafacial plate haired, often with bristly setae along inner margin. Facial ridges haired on lower part. Arista pubescent or short-haired. Palpi blackish brown or light brown. Abdominal tergites with silvery pollinosity anteriorly, shining black at hind margins. A narrow median black stripe present.

The genus is Holarctic in distribution, with 8 species. The larvae live as predators in the nests of bumblebees and social wasps.

### 39. *Brachicoma devia* (Fallén, 1820)
Figs 196-200.

*Tachina devia* Fallén, 1820, Monogr. Musc. Sveciae: 6.
*Brachicoma borealis* Ringdahl, 1932, Notul. ent., 12: 21.

A blackish species with sparse grey pollinosity on thorax, and abdominal tergites silvery grey in anterior 0.5-0.75. Arista pubescent or with very short hairs. Females with 2 proclinate orbital bristles, males without orbitals.

♂. Fronto-orbital and parafacial plates silvery grey pollinose. Frontal vitta black, changing to grey in anterior view. Frons at vertex 0.24-0.27×head-width, frontal vitta narrowest at middle. Fronto-orbital plate more or less densely haired, with strong frontals which continue on to upper part of parafacial plate. No orbital bristles. Parafacial plate with a row of bristly setae along inner margin and some hairs on uppermost part. Antennae black; first antennomere projecting somewhat above lunula, third antennomere 1.7-2.2× as long as second. Arista 1.6-1.8× as long as third antennomere, pubescent or with very short hairs. Vibrissae well-developed, facial ridges setose on lower 0.3-0.5. Palpi and proboscis well-developed, dark brown. Thorax sparse-

93

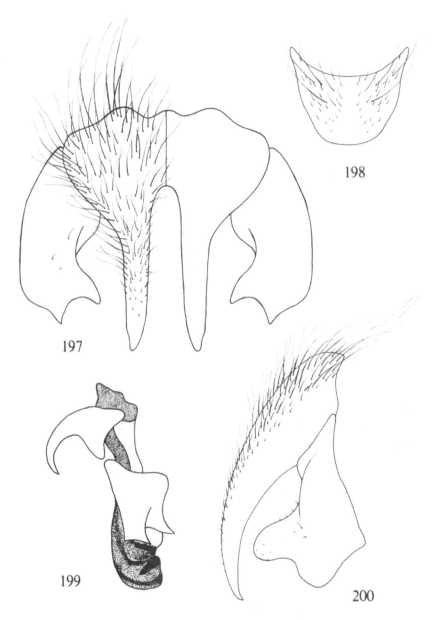

Figs 197-200. *Brachicoma devia* (Fallén). – 197: cerci + surstyli, posterior view; 198: ST5 ♂; 199: aedeagus; 200: cerci + surstyli, lateral view.

ly grey pollinose and with 3 black stripes. No prst acr, katepisternal bristles 2:1. Legs black. Mid femur apically with a comb-like row of short pv bristles. Abdomen black with silvery pollinosity on anterior 0.50-0.75 of each of T3-T5 except for a narrow median non-pollinose stripe. T3 with a pair of median marginal bristles, T4-T5 with a row of marginals. Terminalia black, protruding. Aedeagus short and compact (Fig. 199).

♀. Like the male, but with frons 0.35×head-width at vertex, 2 proclinate orbitals (seldom 3), and without mid femoral pv comb.

Length ♂♀. 6.0-11.0 mm.

Distribution. Very common in Fennoscandia and Denmark wherever suitable hosts occur. Not recorded from the northernmost provinces of Finland. – Widely distributed in the Holarctic region.

Biology. Larvae live in nests of bumblebees and social wasps, preying on the progeny, especially the prepupae (Alford 1975). In Denmark it has been bred from nests of *B. agrorum* (Fabricius), *B. terrestris* (Linnaeus), and *B. soroensis* (Fabricius) (Lundbeck 1927; Schousboe 1981); in Sweden from nests of *B. hypnorum* (Linnaeus), *B. lapidarius* (Linnaeus), and *B. terrestris* (Linnaeus) (Hasselrot 1960); and in Norway from *B. agrorum* (Fabricius) (Rognes 1986a).

Adults may be attracted to decaying meat (Gregor & Povolný 1961).

## Genus *Nyctia* Robineau-Desvoidy, 1830

*Nyctia* Robineau-Desvoidy, 1830, Essai Myod.: 262.
Type species: *Nyctia carceli* Robineau-Desvoidy, 1830, = *Musca halterata* Panzer, 1798.

A monotypic and very characteristic Palaearctic genus. Colour black, pollinosity almost absent. Wings fumose, cell $r_{4+5}$ open, closed, or petiolate. Males with narrow frons, a mid femoral pv comb, and a highly apomorphic aedeagus (Fig. 203). Females with broad frons and proclinate orbital bristles. Abdominal T3 with a pair of strong median marginal bristles.

Some variation in wing morphology exists. Many Mediterranean populations are slightly more pollinose, have a yellow basicosta (normally black), and have a petiolate cell $r_{4+5}$.

### 40. *Nyctia halterata* (Panzer, 1798)
Figs 201-203.

*Musca halterata* Panzer, 1798, Fauna insect. germ., 54: 13.
*Musca maura* Fabricius, 1805, Syst. antl.: 302.

A black species with wings distinctly infuscated along costal margin. Males with narrow frons and without orbital bristles, females with 2 proclinate orbitals. Parafacial plate with a row of bristly setae.

♂. Head black, parafacial plate with thin grey pollinosity. Narrowest part of frons

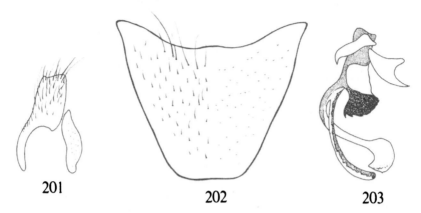

201
202
203

Figs 201-203. *Nyctia halterata* (Panzer). – 201: cerci + surstyli, lateral view; 202: ST5 ♂; 203: aedeagus.

0.1-0.2×head-width. Frontal vitta narrowest just in front of anterior ocellus. Fronto-orbital plate with numerous strong frontals and short hairs. No orbital bristles. Parafacial plate narrow, with a row of bristly setae. Second antennomere with a long bristle just shorter than arista. Third antennomere 1.1-1.3×length of second. Arista plumose. Vibrissae well-developed. Gena with long bristles. Thorax black. Two pairs of prst acr. Katepisternal bristles 1:1. Subscutellum slightly swollen. Legs black, mid femur with a comb-like row of short pv bristles. Wings distinctly infuscated along costal margin and along the veins. Costal spine strong. Cell $r_{4+5}$ open, occasionally closed and more or less petiolate. Abdomen black. T3 with a pair of median marginal bristles, T4-T5 with a row of marginals.

♀. Like the male but less hairy, with frons 0.27-0.30×head-width at vertex, 2 proclinate orbitals, and mid femur without apical comb. Abdominal T6 distinctly visible.

Length ♂♀. 4.0-9.0 mm.

Distribution. Occurring in Denmark: F, but not recorded from Fennoscandia. – Europe, including the British Isles and Canary Islands; western USSR and North Africa.

Biology. Bred from snails (Verves 1982b).

## Genus *Paramacronychia* Brauer & Bergenstamm, 1889

*Paramacronychia* Brauer & Bergenstamm, 1889, Denkschr. Akad. Wiss. Wien, Kl. math.-naturw., 56(1): 116.
Type species: *Macronychia flavipalpis* Girschner, 1881.

Large blackish species; somewhat *Brachicoma*-like. Frons and lower facial margin

96

somewhat protruding. Gena broad, at least 0.5 × eye-height. Females with broad frons and 1-2 proclinate orbital bristles, males with narrower frons and without orbitals. Parafacial plate densely haired. Antennae short, arista bare. Palpi brown to yellow. Wing-cell $r_{4+5}$ open, but occasionally closed at wing-margin. Abdominal tergites silvery pollinose, with a median non-pollinose black stripe.

A single species in the Palaearctic region.

### 41. *Paramacronychia flavipalpis* (Girschner, 1881)
Figs 204, 205.

*Macronychia flavipalpis* Girschner, 1881, Ent. Nachr., 7: 279.
*Paramacronychia hackmani* Verves, 1979, Ann. ent. fenn., 45(1): 31. **Syn.n.**

A blackish species with sparse grey pollinosity on thorax but more extensive grey pollinosity on abdominal tergites. Arista bare. Females with 1-2 proclinate orbital bristles, males without orbitals.

♂. Fronto-orbital and parafacial plates grey pollinose. Frontal vitta black or reddish black. Frons at vertex 0.19 × head-width, frontal vitta narrowest in front of ocellar triangle. Fronto-orbital and parafacial plates densely haired. No orbital bristles. Antennae black; first antennomere slightly projecting above lunula, third antennomere 1.0-1.4 × as long as second. Vibrissae well-developed, facial ridges setose on lower 0.2-0.3. Gena broad, at least 0.5 × eye-height. Palpi brown to yellow. Thorax sparsely grey pollinose and with 3 black stripes. No prst acr, katepisternal bristles 2:1. Legs black. Mid femur with a comb-like row of short pv bristles. Wing-cell $r_{4+5}$ normally open but occasionally closed at wing-margin. Abdomen black with silvery-grey pollinosity on anterior 0.7-0.8 of each of T3-T5 except for a median non-pollinose stripe.

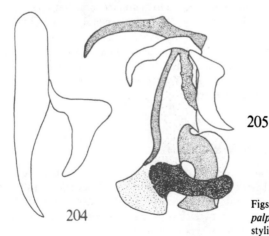

**205**

**204**

Figs 204, 205. *Paramacronychia flavipalpis* (Girschner). – 204: cerci + surstyli, lateral view; 205: aedeagus.

97

T1+2 often with a pair of long median marginal bristles, T3 with 2 pairs of marginals or a complete row, T4-T5 with a row of marginals. Terminalia black, protruding. Aedeagus highly characteristic (Fig. 205).

♀. Like the male in general appearance. Frons much broader, fronto-orbital plate with 1-2 proclinate orbitals, and mid femur without apical pv comb. Abdominal T6 very broad, densely pollinose, and distinctly visible.

Length ♂♀. 7.0-12.0 mm.

Distribution. Only a single specimen is known from Fennoscandia and Denmark: a male with no data except for a small red label, deposited in ZML. The specimen may have been collected by Zetterstedt, who used red labels for specimens collected in Skåne (R. Danielsson *in litt.* 1985). The record needs confirmation. – Widely distributed in the Palaearctic region, from western Europe east to Mongolia and China. Not in the British Isles.

Biology. Unknown.

## SUBFAMILY SARCOPHAGINAE

A very well-defined group of medium-sized (rarely small) to large species. The monophyly of the group is indicated by several apomorphies. Two very distinct ones are the hairs on posterior surface of hind coxae and the freely-exposed abdominal sternites 2-4 in the males. Other characters associated with the group are: female frons broad and with proclinate orbital bristles, males mostly with narrower frons and without proclinate orbitals. Arista generally plumose, parafacial plate with a row of setae on lower half near eye-margin. Occiput and postgena often with white hairs. Notopleuron with 2 strong and 2 weak bristles, and often with additional hairs. Katepimeral bristles 1:1 or 3:1. Katepisternum fused to meron (coxopleural streak absent). Prosternum and postalar walls often haired. Mid tibia generally with 2 ad bristles. Male abdominal ST5 deeply excavated and often with bristles on posterior (inner) margin of each arm. Male terminalia with 2 segments, but without a row of bristles at the point of fusion between T6 and syntergosternite 7+8. Surstyli varying from normal size to rudimentary. Phallotreme on ventral surface of aedeagus. Acrophallus often highly complex. Epiphallus rarely present.

## TRIBE RAVINIINI

Most diversified on the American continent; only a single species in the Old World (apart from a few species in the Pacific archipelagos). Row of frontal bristles not sharply curved outwards at lunula. Male mid femur with an apical comb of short stubby bristles. Aedeagus without articulation between basi- and distiphallus, and acrophallus with an undivided phallotreme. Female hypoproct well sclerotised.

# Genus *Ravinia* Robineau-Desvoidy, 1863

*Ravinia* Robineau-Desvoidy, 1863, Hist. nat. Dipt. Paris, 2: 434.
Type species: *Sarcophaga haematodes* Meigen, 1826, = *Musca pernix* Harris, 1780.

Head with frontal bristles in a straight, or almost straight row. Thorax with distinct prst acr and 3-4 post dc. Wing vein $R_1$ bare or setose; species with setose $R_1$ are often assigned to *Chaetoravinia* Townsend, but this will obviously make the remaining *Ravinia* paraphyletic. Terminalia reddish yellow.

♂. Cerci often broadly separated in posterior view, more seldom closely adpressed. Abdominal ST5 without window, deeply incised and with a brush of short bristles along inner margin.

*Ravinia* is represented by several species on the American continent. Only a single species occurs in the Old World: the Palaearctic and Oriental *R. pernix.*

## 42. *Ravinia pernix* (Harris, 1780)
Figs 206-209.

*Musca pernix* Harris, 1780, Exp. Engl. Ins.: 84.
*Musca striata* Fabricius, 1794, Entom. Syst., 4: 315. Preocc. by Gmelin, 1790.
*Musca haemorrhoidalis* Fallén, 1817, K. svenska VetenskAkad. Handl., [3] 1816: 237.
Preocc. by Villers, 1789.

Easily recognised amongst the Palaearctic Sarcophaginae by the configuration of the

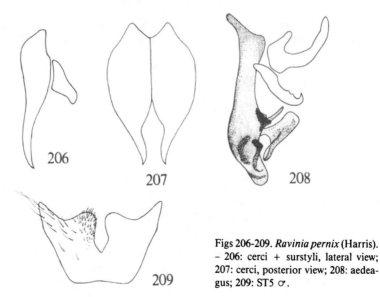

206
207
208
209

Figs 206-209. *Ravinia pernix* (Harris). - 206: cerci + surstyli, lateral view; 207: cerci, posterior view; 208: aedeagus; 209: ST5 ♂.

frontal bristles, the very strong prst acr, the reddish yellow terminalia, and the short costal spine.

♂. Narrowest part of frons 0.23-0.27 × head-width. Row of frontal bristles straight or slightly curved outwards at lunula (Fig. 10). Row of parafacial setae weak. Thorax with 3 post dc and 2-3 strong prst acr. Scutellum without apicals. Legs: mid femur with a row of short av bristles, a row of pv, and a distinct apical pv comb of short, robust bristles. Mid tibia without av bristles. Hind trochanter with numerous long bristly setae. Hind tibia without elongated hairs. Abdominal T3 without median marginal bristles. Terminalia reddish yellow. Protandrial segment grey pollinose. Cerci broadly separated in posterior view (Fig. 207). Aedeagus very characteristic (Fig. 208).

♀. Narrowest part of frons 0.34-0.38 × head-width. Mid femoral organ absent. Mid tibia with a strong av (or v) bristle. Terminalia reddish yellow. T6 entire, with strong marginals.

Length ♂♀. 4.5-8.5 mm.

Distribution. Very common in Denmark and southern and central parts of Fennoscandia. I have been unable to confirm Ringdahl's (1952) record of *R. pernix* from Lu. Lpm. – Widely distributed in the Palaearctic region and temperate + subtropical parts of the Oriental region.

Biology. Breeds in animal and human faeces, the larva probably living partly as a predator (Pickens 1981), but also bred from dead snails (Séguy 1941). Breeding records from various insects need confirmation.

## TRIBE PROTODEXIINI

Frontal bristles sharply curved outwards at lunula. Male cerci with a distinct bend, and aedeagus with an articulation between basi- and distiphallus. Acrophallus with an undivided phallotreme. Female sternites 7-8 fused and modified into a larvipositor. A convincing definition of the group is still needed.

Mostly parasites or parasitoids of insects and myriapods, but some larvae live as predators in pitchers of *Sarracenia*. The Protodexiini is a rather species-rich tribe and has been divided into several genera, especially by American authors. However, until a thorough revision of the American species brings new information on the delimitation of the tribe, it may be better to assign all the species to the genus *Blaesoxipha*.

## Genus *Blaesoxipha* Loew, 1861

*Blaesoxipha* Loew, 1861, Wien. ent. Mschr., 5: 384.
   Type species: *Blaesoxipha grylloctona* Loew, 1861, = *Sarcophaga laticornis* Meigen, 1826.

Medium-sized species. Thorax with 3 post dc and distinct, often strong prst acr. Abdominal tergites with 3 spots which may coalesce into 3 longitudinal stripes. Colour not, or only slightly, changing with the incidence of light.

Male terminalia rather small, almost hidden below T5. Cercal prongs abruptly bent dorsally, often terminating in a small hook. Female often lighter in colour than the male. T6 undivided. ST7-ST8 fused and forming a larvipositor which may be laterally compressed and blade-like, flattened and recurving beneath the abdomen, or more or less trough-like.

Most species are insect parasites or parasitoids, mostly attacking grasshoppers (especially Acrididae), but hosts may include beetles, cicadas, mantids, and even myriapods. Some species, often separated off into the genus *Fletcherimyia*, are predators/scavengers in the proteolytic fluids of *Sarracenia*-cups (their assignment to the Protodexiinae is still debated).

Adults often have a complex larviposition behaviour, and some species are nocturnal or crepuscular. The larvae feed mainly on the fat-body and haemolymph of the host. In acridid hosts, the mature larva often escapes through the dorsal part of the neck-membrane, and if only attacked by a single larva the host may survive, although unable to reproduce. Death may follow due to excessive dehydration through the wound (Wood 1933; Clausen 1940; Léonide & Léonide 1975).

Verves (1985) provides descriptions and keys to all Palaearctic species, and Léonide & Léonide (1986) monograph the French species.

## Key to species of *Blaesoxipha*

1 Males.................................................................... 2
- Females.................................................................. 7
2(1) Mid tibia with one av bristle ...................................... 3
- Mid tibia without av bristles ...................................... 6
3(2) Palpi light brown to yellowish apically. Abdominal T3
without median marginal bristles. Cerci flattened in a
transverse plane (Figs 214, 215). Aedeagus as in Fig. 217 ...............
........................................ 44. *pygmaea* (Zetterstedt)
- Palpi dark brown to blackish throughout. Abdominal T3
with a pair of median marginal bristles. Cerci not flattened
in a transverse plane ............................................. 4
4(3) Cercal prongs strongly bent. Angle between cercal prongs
and cercal base about 130° (Fig. 210). Aedeagus as in Fig.
211.............................. 43. *agrestis* (Robineau-Desvoidy)
- Cercal prongs moderately bent. Angle between cercal
prongs and cercal base about 150° (Fig. 221). Aedeagus as
in Figs 223, 229 ................................................. 5
5(4) Cercal prongs narrow; broadest at base and gradually
tapering (Fig. 221). Cercal bases forming a single plane,
not raised into a slight median keel along the adjoining
margins .............................. 45. *plumicornis* (Zetterstedt)
- Cercal prongs broader and with greatest width just distal
to base (Fig. 227). Adjoining margins of cercal bases

raised into a low but distinct median keel ......... 46. *laticornis* (Meigen)

6(2)   Ventral margin of cerci almost straight (Fig. 237). Aede-
       agus as in Fig. 238, juxta not distinctly arching over the
       acrophallus ............................... 48. *erythrura* (Meigen)
–      Ventral margin of cerci distinctly curved (Fig. 232). Aede-
       agus with juxta arching over the acrophallus (Fig. 236) ................
       ......................................... 47. *rossica* Villeneuve

7(1)   Larvipositor red and recurving beneath the abdomen ................ 8
–      Larvipositor brown and shovel-shaped or straight .................. 9

8(7)   Abdominal ST6 almost as broad as ST5, dorsal concavity
       of larvipositor continuing to apex (Fig. 239) ...... 48. *erythrura* (Meigen)
–      Abdominal ST6 distinctly narrower than ST5, dorsal con-
       cavity of larvipositor not continuing to apex (Fig. 234)..................
       ......................................... 47. *rossica* Villeneuve

9(7)   Larvipositor short, scoop- or shovel-shaped (Figs 212,
       213)............................... 43. *agrestis* (Robineau-Desvoidy)
–      Larvipositor well-developed and more or less pointed ............... 10

10(9)  Larvipositor very large and laterally compressed into a
       sabre-like blade (Fig. 230) ..................... 46. *laticornis* (Meigen)
–      Larvipositor shorter, not sabre-like ........................... 11

11(10) Abdominal T3 with median marginal bristles. Tip of larvi-
       positor blunt and slightly swollen (Figs 225, 226). Palpi
       not or only faintly yellowish apically........ 45. *plumicornis* (Zetterstedt)
–      Abdominal T3 without median marginal bristles. Tip of
       larvipositor gradually tapering in dorsal or ventral view
       (Figs 219, 220). Palpi with distinctly yellow tips . 44. *pygmaea* (Zetterstedt)


## 43. *Blaesoxipha agrestis* (Robineau-Desvoidy, 1863)
   Figs 210-213.

*Listeria agrestis* Robineau-Desvoidy, 1863, Hist. nat. Dipt. Paris, 2: 600.
*Blaesoxipha lineata;* auctt. *nec* Fallén, 1817.

Grey species with distinct black median stripe and lateral spots on abdomen.
      ♂. Similar to *B. plumicornis* in general appearance. Narrowest part of frons about
0.16 × head-width. Abdomen densely grey pollinose with a median black stripe and
lateral black spots which may coalesce into lateral stripes. T3 with a pair of median
marginal bristles. Terminalia: cerci in profile broader and much more bent than in *B.
plumicornis,* angle between cercal bases and cercal prongs about 130° (Fig. 210). Aedeagus as in Fig. 211.
      ♀. Abdominal colour as in male. Larvipositor very short and broad, almost scoop-
or shovel-shaped. ST8 with strong bristles at postero-lateral angles (Figs 212, 213).
      Length ♂♀. 5.0-8.5 mm.

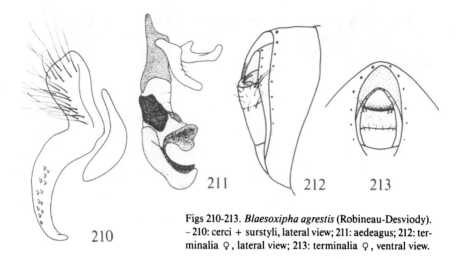

Figs 210-213. *Blaesoxipha agrestis* (Robineau-Desviody). – 210: cerci + surstyli, lateral view; 211: aedeagus; 212: terminalia ♀, lateral view; 213: terminalia ♀, ventral view.

Distribution. Somewhat rare in Denmark and southern Sweden. In Norway from Ø and B. Recorded from Finland by Verves (1986). – Widely distributed in the Palaearctic region, and northern parts of the Afrotropical region. Introduced to Sand Island, Honolulu, for biological control of the locust *Schistocerca nitens* Thunberg, but not established (Chong 1968; Hardy 1981).

Biology. Reared from a broad array of acridid genera: see Wood (1933), Rohdendorf (1937), Léonide & Léonide (1971), and Verves (1985). The females larviposit at random on the host, and the larvae seek and penetrate an intersegmental or arthrodial membrane. Larviposition may even take place on flying hosts.

Note. The identity of *B. lineata* (Fallén) was revised by Pape (1986).

Verves (1986) does not accept the use of *agrestis* Robineau-Desvoidy, 1863 for the present species, and he proposes the name *campestris* Robineau-Desvoidy, 1863, which, however, seems to be a miscitation of *agrestis*.

## 44. *Blaesoxipha pygmaea* (Zetterstedt, 1845)
Figs 214-220.

*Sarcophaga pygmaea* Zetterstedt, 1845, Dipt. Scand., 3: 1302.
*Blaesoxipha berolinensis* Villeneuve, 1912, Annls hist.-nat. Mus. natn. hung., 10: 612.

Greyish species, slightly smaller than other Fennoscandian and Danish species of *Blaesoxipha*. Easily recognisable in both sexes by the light brown to yellow palpal tips.

♂. Narrowest part of frons 0.17-0.18 × head-width. Palpi brown to blackish brown in about proximal 0.6, light brown to yellowish in distal 0.4. Abdomen grey pollinose with 3 stripes, the lateral stripes narrow or indistinct. T3 without median marginal bristles. Terminalia: cerci characteristically flattened in transverse plane; in posterior view with greatest width distal to middle (Fig. 215).

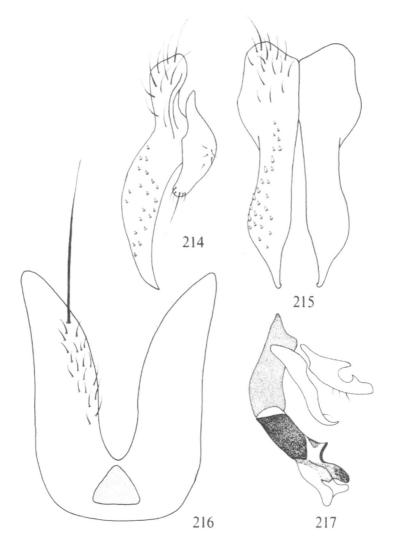

Figs 214-217. *Blaesoxipha pygmaea* (Zetterstedt). – 214: cerci + surstyli, lateral view; 215: cerci, posterior view; 216: ST5 ♂; 217: aedeagus.

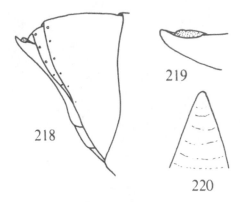

Figs 218-220. *Blaesoxipha pygmaea* (Zetterstedt). – 218: larvipositor, lateral view; 219: tip of larvipositor, lateral view; 220: tip of larvipositor, ventral view.

♀. Lighter grey than male. Larvipositor usually protruding below T6 and gradually tapering in dorsal or ventral view. In lateral view pointed and with a slight concavity at dorso-apical margin (Figs 218, 219).

Length ♂♀. 4.2-7.0 mm.

Distribution. Recorded from eastern Denmark: NEZ and southern Sweden: Sk. where it is somewhat rare or local. No records from Norway or Finland. – Europe, east to Siberia, Mongolia, and China. Oriental region (Pakistan).

Biology. Larvae are injected between the genital appendages of the acridid host. Apparently only female grasshoppers are parasitised (Léonide 1964). Recorded from Acrididae: *Aiolopus, Chorthippus, Locusta, Oedaleus, Schistocerca* (Verves 1985).

### 45. *Blaesoxipha plumicornis* (Zetterstedt, 1859)
Figs 221-226.

*Musca lineata* Fallén, 1817, K. Vetensk.Akad. Handl., [3] 1816: 238. Preocc. by Harris, 1776 and Fabricius, 1781.
*Miltogramma plumicornis* Zetterstedt, 1859, Dipt. Scand., 13: 6153.
*Sarcophaga gladiatrix* Pandellé, 1896, Revue Ent., 15: 205.

Sexual dimorphism distinct: males blackish grey with 3 black stripes on abdomen, females densely grey with a single narrow median stripe on abdomen. Palpi black.

♂. Narrowest part of frons 0.13-0.15 × head-width. Abdomen grey pollinose with 3 black stripes, the lateral stripes narrow or indistinct. T3 with 1-2 pairs of median marginal bristles. Terminalia: cercal bases forming a single plane, not raised into a slight median keel along the adjoining margins. Cercal prongs narrow, gradually tapering in lateral view and at most 1.0-1.5 × width of aedeagal tube distal to basiphallus. Apical hook distinct. Angle between cercal bases and cercal prongs about 150°.

♀. Densely grey pollinose with a narrow median black or brownish black stripe.

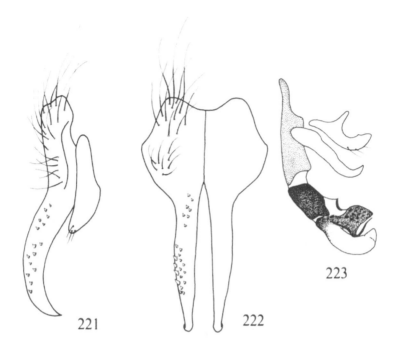

Figs 221-223. *Blaesoxipha plumicornis* (Zetterstedt). – 221: cerci + surstyli, lateral view; 222: cerci, posterior view; 223: aedeagus.

Occasionally traces of lateral stripes present. Larvipositor protruding below T6, but often only with extreme tip. Tip blunt and slightly swollen (Figs 225, 226).

Length ♂♀. 5.0-8.5 mm.

Figs 224-226. *Blaesoxipha plumicornis* (Zetterstedt). – 224: larvipositor, lateral view; 225: tip of larvipositor, lateral view; 226: tip of larvipositor, ventral view.

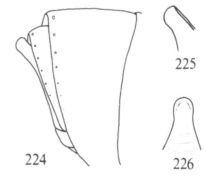

Distribution. Recorded from eastern Denmark, southern and south-eastern Sweden, in Norway north to MR, and in southern Finland. – Palaearctic, from British Isles east to Mongolia, China, and Japan.

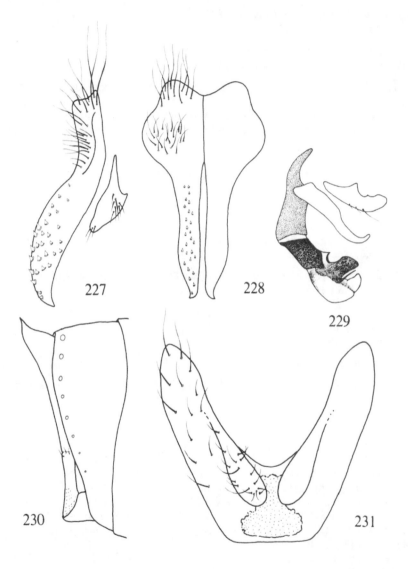

Figs 227-231. *Blaesoxipha laticornis* (Meigen). – 227: cerci + surstyli, lateral view; 228: cerci, posterior view; 229: aedeagus; 230: larvipositor, lateral view; 231: ST5 ♂.

Biology. Bred from a broad array of acridid genera: see Léonide & Léonide (1979) and Verves (1985).


### 46. *Blaesoxipha laticornis* (Meigen, 1826)
  Figs 227-231; pl. 2:2,3.

*Sarcophaga laticornis* Meigen, 1826, Syst. Beschr., 5: 27.
*Blaesoxipha grylloctona* Loew, 1861, Wien. ent. Mschr., 5: 386.

Very similar to *B. plumicornis,* especially in the male sex, but separable by the following characters:
  ♂. Adjoining margins of cercal bases raised into a low but distinct median keel. Cercal prongs in lateral view somewhat broader than *B. plumicornis,* greatest width just distal to base (Fig. 227) and about 1.7-2.0 × width of aedeagal tube distal to basiphallus.
  ♀. Larvipositor laterally compressed into a sabre-like blade, distinctly protruding behind (Fig. 230).
  Length ♂♀. 5.0-8.5 mm.

Distribution. Common in Denmark and South Sweden, but not recorded from Norway or Finland. – Palaearctic, from western Europe east to China, Mongolia, and Japan. Not in the British Isles.

Biology. Bred from Acrididae: *Chorthippus, Omocestus* (Richards & Waloff 1948; Parmenter 1950; Léonide & Léonide 1982).


### 47. *Blaesoxipha rossica* Villeneuve, 1912
  Figs 232-236.

*Blaesoxipha rossica* Villeneuve, 1912a, Annls hist.-nat. Mus. natn. hung., 10: 611.

Blackish grey species with yellowish red terminalia. Abdominal T3 with weak median marginal bristles, seldom without (as noted by Emden (1954)).
  ♂. Narrowest part of frons 0.17-0.20 × head-width. Mid tibia without av bristles. Abdomen black with silvery grey pollinosity. T3-T5 with 3 distinct black stripes, occasionally with the lateral stripes broken up into rectangular spots. T3 with or without median marginal bristles, but when present they are often somewhat weak or asymmetrically developed. Terminalia: cerci with ventral margin distinctly curved (Fig. 232). Aedeagus with juxta arching over the acrophallus (Fig. 236).
  ♀. Colour as in male. Mid tibia always with a strong av bristle. Terminalia yellowish red. T6 entire. ST6 distinctly narrower than ST5 (Fig. 234). ST8 forming a recurved, pointed larvipositor. Tip of larvipositor flat, dorsal concavity not continuing to apex (Fig. 234).
  Length ♂♀. 6.0-9.5 mm.

Distribution. Not rare in Denmark and southern Sweden. Misidentified as *B. fos-*

*soria* (Pandellé) in Ringdahl (1952). From Norway a single female from MR. Not recorded from Finland. – Palaearctic, east to Siberia, Mongolia, and China.

Biology. Bred from Acrididae: *Chorthippus, Euchorthippus, Gomphocerus.* The female injects the larvae through the intersegmental membranes of the host abdomen (Léonide 1967; Léonide & Léonide 1971).

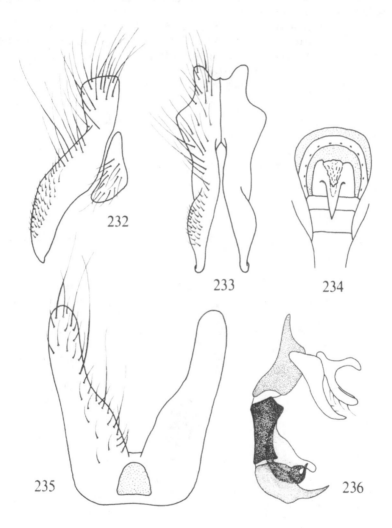

Figs 232-236. *Blaesoxipha rossica* Villeneuve. – 232: cerci + surstyli, lateral view; 233: cerci, posterior view; 234: larvipositor, ventral view; 235: ST5 ♂; 236: aedeagus.

**48. *Blaesoxipha erythrura* (Meigen, 1826)**
  Figs 237-239.

*Sarcophaga erythrura* Meigen, 1826, Syst. Beschr., 5: 30.

Very similar to *B. rossica* in general appearance, but always with a strong and well-defined pair of median marginal bristles on abdominal T3. Separable from *B. rossica* by the following characters:

  ♂. Narrowest part of frons 0.14-0.16 × head-width. Terminalia: cerci with ventral margin almost straight, and tip rather blunt (Fig. 237). Aedeagus with a pair of partly membraneous lappets (?harpes), and juxta shorter than in *B. rossica* (Fig. 238).

  ♀. ST6 almost as broad as ST5 (Fig. 239). Larvipositor shorter than in *B. rossica,* with the dorsal concavity continuing to apex (Fig. 239).

  Length ♂♀. 6.0-9.5 mm.

Distribution. No records from Denmark and Norway (note that Lundbeck (1927) misidentified Danish specimens of *B. rossica* as the present species, as was correctly stated by Lyneborg (1960)). Rare in southern Sweden: Sm., Öl., but common in Finland north to ObN. – Western Palaearctic, east to Siberia, Mongolia, and China.

Biology. Bred from Acrididae: *Chorthippus, Chrysochraon, Omocestus.*

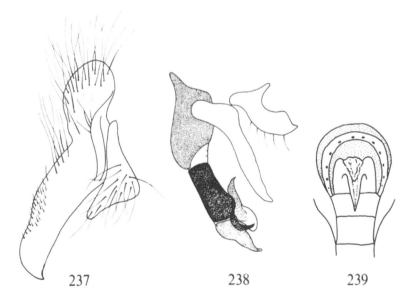

Figs 237-239. *Blaesoxipha erythrura* (Meigen). – 237: cerci + surstyli, lateral view; 238: aedeagus; 239: larvipositor, ventral view.

110

# TRIBE SARCOPHAGINI

Frontal bristles sharply curved outwards at lunula. Aedeagus with articulation between basi- and distiphallus. Acrophallus with a division of the phallotreme into two parts.

Fennoscandian and Danish species of Sarcophagini are extremely similar in general appearance, all being more or less greyish black with a tessellate abdominal pattern which changes with the incidence of light (pl. 2:4).

Generic delimitations within the Sarcophagini are mostly based on similarities in the male and female terminalia, and some future rearrangement may be expected. A confusing number of genus-group names have been created, which are of limited value to anyone but the specialist. However, the lumping-together of all the species of the Sarcophagini into a single genus, *Sarcophaga* sensu lato, would immediately create the need for subdivisions (subgenera or species-groups) in order to find a way of handling such a huge genus. I have decided to use the generic names in current use (at least by Palaearctic authors), but to avoid a key to genera which would be difficult to use without loosing one's way amongst tortuous descriptions of aedeagal appendages and peculiarities.

Species keys have been constructed separately for males and females. Males of most species can be identified solely on external characters (visible without dissection of terminalia), but the diagnostic characters are variable, and in many instances only the copulatory apparatus will provide a definite identification. The best method of identifying male specimens of Sarcophagini is to compare their terminalia directly with the figures, and then – if necessary – to go back and check with the key.

Females are much more problematic to identify, and they are usually entirely neglected in existing keys (Lundbeck 1927; Rohdendorf 1937; Kano *et al.* 1967; Mihályi 1979). The key to females of the British species of Sarcophagini published by Day & Assis Fonseca (1955) is rather difficult to use and suffers from a total lack of illustrations. The key of Emden (1954) is much better, but the small number of species covered makes it of limited value outside the British Isles. The present key may be difficult to use without previous knowledge of the group as some of the diagnostic characters may be hard to detect (e.g. the mid femoral organ), others are rather subtle (e.g. the thickness of the marginal bristles on abdominal T6), whilst others are used although they may be somewhat unreliable (e.g. presence/absence of median marginal bristles on abdominal T3). Future research may add to our knowledge of infraspecific variability. More numerous (and more accurate) drawings of the female terminalia may become available, and new and more reliable characters may be used. The female internal genitalia may provide several characters of value for identification purposes. The complex male distiphallus is generally thought to reflect the topography of the female genital ducts, and the signum (a sclerotisation in the dorsal wall of the uterus) or the spermathecae may provide useful characters.

111

# Key to species of Sarcophagini

**Males**

1    3 strong post dc bristles (Fig. 7), first bristle always closer
to suture than to second ........................................... 2

–    4-6 post dc bristles (Figs 8,9), the anterior 1-3 may be rather
weak and the first strong bristle then distinctly closer to
second than to suture ............................................. 31

2(1)   Epandrium red or yellowish ....................................... 3

–    Epandrium black or brownish black ............................... 8

3(2)   Scutellar apicals absent ............... 50. *Discachaeta pumila* (Meigen)

–    Scutellar apicals present ......................................... 4

4(3)   Abdominal T3 with a pair of median marginal bristles ................ 5

–    Abdominal T3 without median marginal bristles ..................... 7

5(4)   Protandrial segment with a pollinose spot which is distinctly longer than broad. Cerci with a very small indentation
at apex (Fig. 298). Aedeagal membrane below vesica
greatly swollen (Fig. 301). ............. 61. *Heteronychia vagans* (Meigen)

–    Protandrial segment with a pollinose spot which is circular
or distinctly broader than long. Cerci with apex evenly
rounded. Aedeagal membrane below vesica only slightly
swollen ........................................................... 6

6(5)   Hind femoral av bristles often hair-like. Apical part of juxta distinctly longer than juxtal appendages (Fig. 293) .................
............................. 59. *Heteronychia haemorrhoa* (Meigen)

–    Hind femoral av bristles never hair-like. Apical part of
juxta shorter than juxtal appendages (Fig. 296) ......................
...................... 60. *Heteronychia boettcheriana* (Rohdendorf)*

7(4)   Hind femur with a distinct row of bristles. Aedeagus as in
Fig. 310. ........................ 63. *Heteronychia proxima* (Rondani)

–    Hind femur without av bristles. Aedeagus as in Fig. 306 ..............
................................. 62. *Heteronychia vicina* (Macquart)

8(2)   Wing vein R₁ setose ............................................... 9

–    Wing vein R₁ bare ................................................ 11

9(8)   Postgenal hairs mostly black, with only a few white hairs in
posterior part. Scutellar apicals well-developed. Hind trochanter with long setae ventromedially. Protandrial segment without marginal bristles. Aedeagus as in Fig. 312
............................... 64. *Pierretia sexpunctata* (Fabricius)

–    Postgenal hairs white. Scutellar apicals weak or absent.
Hind trochanter with short bristly setae ventromedially.

---

*Specimens of *Heteronychia proxima* with median marginal bristles on T3 key out here, but
they are easily recognised by the aedeagal structure (Fig. 310).

112

Protandrial segment with marginal bristles . . . . . . . . . . . . . . . . . . . . . . . . . 10
10(9)  Wing vein R₁ with 4-7 setae. Aedeagus as in Fig. 283 . . . . . . . . . . . . . . . . . .

. . . . . . . . . . . . . . . . . . . . . . . 57. *Heteronychia depressifrons* (Zetterstedt)
–      Wing vein R₁ with 1-3 setae. Aedeagus as in Fig. 288 . . . . . . . . . . . . . . . . . .

. . . . . . . . . . . . . . . . . . . . . . . . . . . 58. *Heteronychia bezziana* (Böttcher)
11(8)  Postgenal hairs mostly black, with only a few white hairs in
       posterior part. Aedeagus as in Fig. 312 64. *Pierretia sexpunctata* (Fabricius)
–      Postgenal hairs white . . . . . . . . . . . . . . . . . . . . . . . . . . . . . . . . . . . . . . . . . . . 12
12(11) Mid femur with an oblong patch of golden hairs on distal
       anterior surface, and sometimes some golden hairs on mid-
       dle part of posterior surface. Aedeagus as in Fig. 338 . . . . . . . . . . . . . . . . . .

. . . . . . . . . . . . . . . . . . . . . . . . . . . 70. *Sarcotachinella sinuata* (Meigen)
–      Mid femur without golden hairs . . . . . . . . . . . . . . . . . . . . . . . . . . . . . . . . . . 13
13(12) 1-4 pairs of prst acr present . . . . . . . . . . . . . . . . . . . . . . . . . . . . . . . . . . . . . . 14
–      No distinct prst acr among the hairs . . . . . . . . . . . . . . . . . . . . . . . . . . . . . 20
14(13) Protandrial segment pollinose . . . . . . . . . . . . . . . . . . . . . . . . . . . . . . . . . . . . 15
–      Protandrial segment non-pollinose and shining . . . . . . . . . . . . . . . . . . . . . 17
15(14) Abdominal T3 without median marginal bristles. Cerci
       almost triangular in profile (Fig. 276). Aedeagus as in Fig. 280 . . . . . . . . . .

. . . . . . . . . . . . . . . . . . . . . . . . . . . 56. *Helicophagella hirticrus* (Pandellé)
–      Abdominal T3 with a pair of median marginal bristles . . . . . . . . . . . . . . . 16
16(15) Abdominal ST3-ST4 with long, erect hairs along lateral
       margins, longest hairs about as long as length of one sterni-
       te. Aedeagus as in Fig. 264 . . . . . . . . . . 53. *Helicophagella agnata* (Rondani)
–      Abdominal ST3-ST4 with short, decumbent hairs along la-
       teral margins, longest hairs about 0.5 × as long as length of
       one sternite. Aedeagus as in Fig. 318 . . . . . 65. *Pierretia nemoralis* (Kramer)
17(14) Scutellar apicals absent. Aedeagus as in Fig. 323 . . . . . . . . . . . . . . . . . . . . . .

. . . . . . . . . . . . . . . . . . . . . . . . . . . 66. *Pierretia villeneuvei* (Böttcher)
–      Scutellar apicals present, although often weak . . . . . . . . . . . . . . . . . . . . . . 18
18(17) Hind tibia with long and dense hairs on pv and v surfaces,
       longest hairs often with wavy tips. Aedeagus as in Fig. 333 . . . . . . . . . . . . .

. . . . . . . . . . . . . . . . . . . . . . . . . . . . . . 69. *Pierretia soror* (Rondani)
–      Hind tibia without long hairs, or with a single row of pv
       hairs without wavy tips . . . . . . . . . . . . . . . . . . . . . . . . . . . . . . . . . . . . . . . . 19
19(18) Posterior part of gena with 3-8 white hairs. Frons 0.27-
       0.33 × head-width. Longest aristal hairs 1.5-3.0 × as long
       as second aristomere. Aedeagus as in Fig. 331 . . . . . . . . . . . . . . . . . . . . . . .

. . . . . . . . . . . . . . . . . . . . . . . . . . . 68. *Pierretia nigriventris* (Meigen)
–      Genal hairs black, white postgenal hairs never extending
       on to posterior part of gena. Frons 0.23-0.28 × head-width.
       Longest aristal hairs 3.0-4.5 × as long as second aristomere.
       Aedeagus as in Fig. 327 . . . . . . . . . . . . . . . . 67. *Pierretia socrus* (Rondani)
20(13) Scutellar apicals absent. Protandrial segment rather slen-

der, with marginal bristles and a broad pollinose spot posteriorly. Aedeagus as in Fig. 248 . . . . . . . . 50. *Discachaeta pumila* (Meigen)
–   Scutellar apicals present . . . . . . . . . . . . . . . . . . . . . . . . . . . . . . . . . . . . . . . . 21
21(20) Abdominal T3 with a pair of median marginal bristles . . . . . . . . . . . . . . 22
–   Abdominal T3 without median marginals . . . . . . . . . . . . . . . . . . . . . . . . 25
22(21) Mid femur without pv bristles. Aedeagus as in Fig. 268 . . . . . . . . . . . . . . .
. . . . . . . . . . . . . . . . . . . . . . . . . . . . . . 54. *Helicophagella rosellei* (Böttcher)
–   Mid femur with at least a few distinct pv bristles . . . . . . . . . . . . . . . . . . . . 23
23(22) Protandrial segment without marginal bristles. Aedeagus
as in Fig. 318 . . . . . . . . . . . . . . . . . . . . . . . 65. *Pierretia nemoralis* (Kramer)
–   Protandrial segment with weak marginals. Aedeagus dif-
ferent . . . . . . . . . . . . . . . . . . . . . . . . . . . . . . . . . . . . . . . . . . . . . . . . . . . . . . 24
24(23) Frons narrow, 0.15-0.18 × head-width. Thorax often with
a few prst acr indicated. Hind trochanter with apical bristle
about 1.3 × length of ventromedian hairs. Aedeagus as in
Fig. 264 . . . . . . . . . . . . . . . . . . . . . . . . 53. *Helicophagella agnata* (Rondani)
–   Frons broader, 0.18-0.24 × head-width. No prst acr indica-
ted. Hind trochanter with apical bristle about 2.0 × length
of ventromedian hairs. Aedeagus as in Fig. 259 . . . . . . . . . . . . . . . . . . . . .
. . . . . . . . . . . . . . . . . . . . . . . . . . 52. *Helicophagella crassimargo* (Pandellé)
25(21) Frons very broad, 0.26-0.29 × head-width. Protandrial
segment non-pollinose or with only weak traces of pollino-
sity. Aedeagus as in Figs 252, 253 . . . 51. *Helicophagella melanura* (Meigen)
–   Frons narrower, less than 0.25 × head-width. Protandrial
segment with distinct pollinosity . . . . . . . . . . . . . . . . . . . . . . . . . . . . . . . . . 26
26(25) Mid femur with at least a few pv bristles . . . . . . . . . . . . . . . . . . . . . . . . . . 27
–   Mid femur without pv bristles . . . . . . . . . . . . . . . . . . . . . . . . . . . . . . . . . . . 30
27(26) Protandrial segment without marginal bristles . . . . . . . . . . . . . . . . . . . . . 28
–   Protandrial segment with marginal bristles, or at least with
bristly hairs on posterior margin . . . . . . . . . . . . . . . . . . . . . . . . . . . . . . . . . 29
28(27) Spinose lobes of vesica shorter than median lobes. Median
processes evenly curved. Apical part of juxta about as long
as basal width (Fig. 343) . . . . . . . . 71. *Thyrsocnema incisilobata* (Pandellé)
–   Spinose lobes of vesica at least as long as median lobes.
Median processes with a subapical bend. Apical part of
juxta distinctly shorter than basal width (Fig. 348) . . . . . . . . . . . . . . . . . . . .
. . . . . . . . . . . . . . . . . . . . . . . . . . 72. *Thyrsocnema kentejana* Rohdendorf
29(27) Epandrium with some long hairs which are as strong as and
longer than marginal bristles of protandrial segment.
Aedeagus as in Fig. 259 . . . . . . . 52. *Helicophagella crassimargo* (Pandellé)
–   All epandrial hairs shorter and weaker than marginal bris-
tles of protandrial segment. Aedeagus as in Fig. 242 . . . . . . . . . . . . . . . . . .
. . . . . . . . . . . . . . . . . . . . . . . . . . . . . 49. *Bellieriomima subulata* (Pandellé)
30(26) Protandrial segment without marginal bristles. Aedeagus

114

as in Fig. 280 . . . . . . . . . . . . . . . . . . . 56. *Helicophagella hirticrus* (Pandellé)
- Protandrial segment with marginal bristles. Aedeagus as in Fig. 274 . . . . . .
. . . . . . . . . . . . . . . . . . . . . . . . . . . . . 55. *Helicophagella noverca* (Rondani)
31(1) Epandrium red or yellowish . . . . . . . . . . . . . . . . . . . . . . . . . . . . . . . . . . . . . 32
- Epandrium black or brownish black . . . . . . . . . . . . . . . . . . . . . . . . . . . . . 34
32(31) Prsc acr absent or hair-like. Protandrial segment most
often with 1-2 pairs of strong marginals. Cerci broadly se-
parated and with a distinct notch at base (Fig. 350) . . . . . . . . . . . . . . . . . .
. . . . . . . . . . . . . . . . . . . . . . . . . . . . . . . . 73. *Bercaea cruentata* (Meigen)
- Prsc acr present and strong. Protandrial segment without
or with weak marginals. Cerci different . . . . . . . . . . . . . . . . . . . . . . . . . . 33
33(32) Protandrial segment with marginal bristles more or less de-
veloped. White hairs of gena and occiput very long and
dense. Surstyli thickened along basal margin (Fig. 407).
Aedeagus as in Fig. 409 . . . . . . . . . . . . . . . . . . . . . . . . . . . . . . . . . . . . . . .
. . . . . . . . . . . . . . . . 83. *Parasarcophaga argyrostoma* (Robineau-Desvoidy)
- Protandrial segment without marginal bristles. White hairs
of gena not very long and dense. Surstyli not thickened
along basal margin. Aedeagus as in Fig. 381 . . . . . . . . . . . . . . . . . . . . . . .
. . . . . . . . . . . . . . . . . . . . . . . . . 78. *Parasarcophaga jacobsoni* Rohdendorf
34(31) Protandrial segment with marginal bristles . . . . . . . . . . . . . . . . . . . . . . . . 35
- Protandrial segment without marginal bristles . . . . . . . . . . . . . . . . . . . . . . 39
35(34) Blackish species. Abdominal T3 without marginal bris-
tles. Epandrium flattened dorsally or with a slight concavity. Ventral abdominal hairs long and dense, about as long
as greatest diameter of hind femur. Aedeagus as in Fig.
377 . . . . . . . . . . . . . . . . . . . . . 77. *Parasarcophaga caerulescens* (Zetterstedt)
- Abdominal T3 with, sometimes without, a pair of median
marginal bristles. Epandrium convex dorsally. Ventral
abdominal hairs shorter, only a few as long as greatest dia-
meter of hind femur . . . . . . . . . . . . . . . . . . . . . . . . . . . . . . . . . . . . . . . . . 36
36(35) Ventral plates of aedeagus smaller than vesica . . . . . . . . . . . . . . . . . . . . . 37
- Ventral plates of aedeagus as large as or larger than vesica . . . . . . . . . . . . 38
37(36) Cercal tip distinctly behind anterior cercal margin (Fig.
416). Vesica 3.0 × as long as height of ventral plates. Juxta
membraneous, more or less enveloping the styli (Fig. 422) . . . . . . . . . . . . . .
. . . . . . . . . . . . . . . . . . . . . . . . . . . . 87. *Sarcophaga subvicina* Rohdendorf
- Cercal tip level with, or slightly projecting beyond, anterior
cercal margin; subapical swelling moderately developed
(Fig. 415). Vesica 1.5 × as long as height of ventral plates.
Juxta sclerotised, not enveloping the styli (Fig. 421) . . . . . . . . . . . . . . . . . .
. . . . . . . . . . . . . . . . . . . . . . . . . . . . . . 84. *Sarcophaga carnaria* (Linnaeus)
38(36) Cercal hook projecting beyond anterior cercal margin;
subapical swelling moderately developed (Fig. 413). Styli

115

reaching beyond margin of ventral plates, styli and juxta
forming an angle of 120-130° with longitudinal axis
of aedeagus (Fig. 419) .............. 86. *Sarcophaga variegata* (Scopoli)
–   Cercal tip level with, or slightly projecting beyond, anterior
cercal margin; subapical swelling well-developed (Fig.
414). Styli not reaching beyond margin of ventral plates,
styli and juxta forming an angle of 95-105° with longitudinal axis of aedeagus (Fig. 420)........ 85. *Sarcophaga lasiostyla* Macquart
39(34) Gena with only black hairs; white hairs restricted to postgena and posterior parts of occiput ................................. 40
–   Gena with at least some white hairs on posterior part ................. 45
40(39) Abdominal ST5 raised basally into a distinct median tooth-
like projection (Fig. 368) .......... 75. *Parasarcophaga aratrix* (Pandellé)
–   Abdominal ST5 with a convex, often more or less keeled
base, but never tooth-like ........................................ 41
41(40) Hind tibia with some slightly elongated pv hairs .................... 42
–   Hind tibia with long dense hairs on both pv and v surfaces ............ 43
42(41) Each arm of abdominal ST5 with marginal bristles and a
tuft of short, stout bristles at base (Fig. 403). Aedeagus as
in Fig. 402........................ 82. *Parasarcophaga similis* (Meade)
–   Each arm of abdominal ST5 with long marginal bristles,
but without a tuft of short, stout bristles at base. Aedeagus
as in Fig. 385 .................. 79. *Parasarcophaga emdeni* Rohdendorf
43(41) Prst acr distinct or at least indicated. Cerci excavated dorsally. Aedeagus with bilobed juxta but without slender juxtal arms (Fig. 372) .............. 76. *Parasarcophaga uliginosa* (Kramer)
–   Prst acr not differentiated among the hairs. Cerci not excavated dorsally. Aedeagus with undivided juxta and slender
juxtal arms...................................................... 44
44(43) Base of abdominal ST5 distinctly keeled, in profile forming an obtuse angle (Fig. 391). Gonopod with an apical
hook. Aedeagus with styli terminating well before apex of
juxtal arms. Apical part of juxta short (Fig. 389). .....................
...................... 80. *Parasarcophaga portschinskyi* Rohdendorf
–   Base of abdominal ST5 convex, but without a distinct keel;
in profile forming almost a right angle (Fig. 399). Gonopod gradually tapering. Aedeagus with styli terminating at
about apex of juxtal arms. Apical part of juxta long (Fig.
396) ........................ 81. *Parasarcophaga pleskei* Rohdendorf
45(39) Hind tibia with a single row of slightly elongated pv hairs.
Aedeagus as in Fig. 385 ......... 79. *Parasarcophaga emdeni* Rohdendorf
–   Hind tibia with long dense hairs on both pv and v surfaces ........... 46
46(45) Gena with white hairs on about posterior 0.4-0.7. Aedeagus as in Fig. 359 ................ 74. *Parasarcophaga albiceps* (Meigen)

–      Gena with white hairs on about posterior 0.1-0.2 . . . . . . . . . . . . . . . . . . . . . 44


## Females

| | | |
|---|---|---|
| 1 | 3 strong post dc bristles (Fig. 7), first bristle always closer to suture than to second . . . . . . . . . . . . . . . . . . . . . . . . . . . . . . . . . . . . . . . . . . . . . . . . . . . 2 | |
| – | 4-6 post dc bristles (Figs 8, 9), the anterior 1-3 may be rather weak and the first strong bristle then distinctly closer to second bristle than to suture . . . . . . . . . . . . . . . . . . . . . . . . . . . . . . . . . 25 | |

2(1)  Terminalia red or yellowish . . . . . . . . . . . . . . . . . . . . . . . . . . . . . . . . . . . . 3
–      Terminalia black or blackish brown . . . . . . . . . . . . . . . . . . . . . . . . . . . . . 7
3(2)  Abdominal T3 with a pair of median marginals . . . . . . . . . . . . . . . . . . . . 4
–      Abdominal T3 without median marginals . . . . . . . . . . . . . . . . . . . . . . . . . 6
4(3)  Abdominal T6 slightly desclerotised and somewhat angular on dorsal median line (Fig. 300) . . . . 61. *Heteronychia vagans* (Meigen)
–      Abdominal T6 fully sclerotised and evenly curved (not angular) on dorsal median line (Fig. 290) . . . . . . . . . . . . . . . . . . . . . . . . . 5
5(4)  Vein $R_1$ setose . . . . . . . . . . . . . . . . 59. *Heteronychia haemorrhoa* (Meigen), and 60. *Heteronychia boettcheriana* (Rohdendorf)
–      Vein $R_1$ bare . . . . . . . . . . . . . . . . . . . . 63. *Heteronychia proxima* (Rondani)
6(3)  Mid femoral organ apical (Fig. 303) or indistinct. Abdominal T6 with strong and dense marginal bristles, the distance between bristles about 1-3 × thickness of a single bristle . . . . . . . . . . . . . . . . . . . . . . . . . . 62. *Heteronychia vicina* (Macquart), and 63. *Heteronychia proxima* (Rondani)
–      Mid femoral organ at middle (Fig. 345). Abdominal T6 with strong, widely-separated bristles and long hairs . . . . . . . . . . . . . . . . . . . . . . . . . . . . . . . . . . . . . . . . . . . . . . . . . . . . . . . . 71-72. *Thyrsocnema* spp.
7(2)  Vein $R_1$ setose . . . . . . . . . . . . . . . . . . . . . . . . . . . . . . . . . . . . . . . . . . . . . 8
–      Vein $R_1$ bare . . . . . . . . . . . . . . . . . . . . . . . . . . . . . . . . . . . . . . . . . . . . . 10
8(7)  Postgenal hairs mostly black, with only a few white hairs in posterior part around occipital foramen . . . . . . . . . . . . . . . . . . . . . . . . . . . . . . . . . . . . . . . . . . . . . . . . . . . . . . 64. *Pierretia sexpunctata* (Fabricius)
–      Postgenal hairs white . . . . . . . . . . . . . . . . . . . . . . . . . . . . . . . . . . . . . . . . 9
9(8)  Vein $R_1$ with 4-7 setae . . . . . . 57. *Heteronychia depressifrons* (Zetterstedt)
–      Vein $R_1$ with 1-3 setae . . . . . . . . . . . . 58. *Heteronychia bezziana* (Böttcher)
10(7) Postgenal hairs mostly black, with only a few white hairs in posterior part around occipital foramen . . . . . . . . . . . . . . . . . . . . . . . . . . . . . . . . . . . . . . . . . . . . . . . . . . . . . . 64. *Pierretia sexpunctata* (Fabricius)
–      Postgenal hairs white . . . . . . . . . . . . . . . . . . . . . . . . . . . . . . . . . . . . . . . 11
11(10) Mid femur with an oblong patch of golden hairs on distal anterior surface, and sometimes some golden hairs on middle part of posterior surface. Mid femoral organ ra-

ther small and situated at middle (Fig. 339) ..........................
................................ 70. *Sarcotachinella sinuata* (Meigen)
       –   Mid femur without golden hairs ................................. 12
12(11) Some pairs of prst acr present. Mid femoral organ absent ........... 13
       –   No prst acr differentiated, or if distinct then mid femoral
          organ large and red .............................................. 16
13(12) Prsc acr well-developed ............ 53. *Helicophagella agnata* (Rondani)
       –   Prsc acr absent or very weakly developed ........................ 14
14(13) Posterior part of gena with 3-8 white hairs close to post-
          gena ........................... 68. *Pierretia nigriventris* (Meigen),
                  and 69. *Pierretia soror* (Rondani)
       –   Posterior part of gena never with white hairs close to
          postgena ...................................................... 15
15(14) Longest aristal hairs 1.5-3.0 × as long as second aristomere .............
................................ 66. *Pierretia villeneuvei* (Böttcher)
       –   Longest aristal hairs 3.0-4.5 × as long as second aristo-
          mere ................................ 67. *Pierretia socrus* (Rondani)
16(12) Abdominal T3 with a pair of median marginal bristles ............... 17
       –   Abdominal T3 without median marginal bristles ................... 21
17(16) Mid femoral organ absent ....................................... 18
       –   Mid femoral organ present....................................... 19
18(17) Abdominal T6 undivided but slightly indented dorsally
          (Fig. 246). Prst acr often weak ......... 50. *Discachaeta pumila* (Meigen)
       –   Abdominal T6 distinctly divided into two sclerites sepa-
          rated by a narrow membraneous strip. Prst acr often
          well-developed .................. 53. *Helicophagella agnata* (Rondani)
19(17) Mid femoral organ in distal 0.3 ......... 65. *Pierretia nemoralis* (Kramer)
       –   Mid femoral organ at middle..................................... 20
20(19) Mid femoral organ often bright red. Marginal bristles of
          T6 0.7-0.8 × as thick as marginals of T5 ...........................
................................ 54. *Helicophagella rosellei* (Böttcher)
       –   Mid femoral organ brown, not bright red. Marginal brist-
          les of abdominal T6 0.3-0.5 × as thick as marginals of T5 ..............
........................ 52. *Helicophagella crassimargo* (Pandellé)
21(16) Mid femoral organ present...................................... 22
       –   Mid femoral organ absent ...................................... 23
22(21) Mid femoral organ in distal 0.5 (Fig. 278), often bright
          red, and of medium size to large... 55. *Helicophagella noverca* (Rondani),
                and 56. *Helicophagella hirticrus* (Pandellé)
       –   Mid femoral organ at middle (Fig. 345), brownish or indi-
          stinct, and of medium size .................. 71-72. *Thyrsocnema* spp.
23(21) Narrowest part of frons 0.35-0.39 × head-width. Termi-
          nalia distinctly protruding and densely pollinose. T6 di-
          vided; membrane between and posterior to the halves

black and sclerotised, forming a pit-like depression (Fig.
255) . . . . . . . . . . . . . . . . . . . . . . . . . 51. *Helicophagella melanura* (Meigen)
- Narrowest part of frons less than 0.35 × head-width. Ter-
minalia at most slightly protruding. Membrane posterior
to T6 never black and sclerotised . . . . . . . . . . . . . . . . . . . . . . . . . . . . . 24
24(23) Abdominal T6 divided into broadly separated halves
(Fig. 244). Lower facial margin not projecting beyond
antennal insertion . . . . . . . . . . . . . . . 49. *Bellieriomima subulata* (Pandellé)
- Abdominal T6 undivided but with a dorsal indentation
(Fig. 246). Lower facial margin slightly projecting
beyond antennal insertion . . . . . . . . . . . . 50. *Discachaeta pumila* (Meigen)
25(1) Terminalia distinctly red or yellowish. Abdominal T3
without median marginal bristles. (Note that females of
*Sarcophaga* spp. often have blackish red terminalia) . . . . . . . . . . . . . . . . 26
- Terminalia black or dark blackish red. T3 with or with-
out median marginal bristles . . . . . . . . . . . . . . . . . . . . . . . . . . . . . . . . . 29
26(25) Abdominal T6 deeply incised on dorsal median line (Fig.
356). ST7 convex (Fig. 355) . . . . . . . . . . . . . 73. *Bercaea cruentata* (Meigen)
- Abdominal T6 evenly curved, at most with a slight me-
dian indentation or a median longitudinal fold. ST7 dif-
ferent . . . . . . . . . . . . . . . . . . . . . . . . . . . . . . . . . . . . . . . . . . . . . . . . . . . . 27
27(26) Marginal bristles of abdominal T6 hair-like, about 0.2-
0.4 × as thick as marginals of T5. T8 reduced. ST7 with a
median knob (Fig. 412). Gena with long dense white
hairs along almost its entire length . . . . . . . . . . . . . . . . . . . . . . . . . . . . . .
. . . . . . . . . . . . . . . 83. *Parasarcophaga argyrostoma* (Robineau-Desvoidy)
- Marginal bristles of abdominal T6 strong, about 0.5-0.7 ×
as thick as marginals of T5. T8 entire, slightly arching
over the cerci (Fig. 393). ST7 without a median knob.
Gena with black or white hairs . . . . . . . . . . . . . . . . . . . . . . . . . . . . . . . . . 28
28(27) Genal hairs white on at least posterior half . . . . . . . . . . . . . . . . . . . . . . . . .
. . . . . . . . . . . . . . . . . . . . . . . . . 78. *Parasarcophaga jacobsoni* Rohdendorf
- Genal hairs black, occasionally with a few white hairs
close to postgena . . . . . . . . 80. *Parasarcophaga portschinskyi* Rohdendorf
(Note that 81. *Parasarcophaga pleskei* Rohdendorf may key out here).
29(25) Terminalia distinctly protruding. T6 divided dorsally,
with closely touching margins (Fig. 374). Blackish spe-
cies . . . . . . . . . . . . . . . . . . . . 77. *Parasarcophaga caerulescens* (Zetterstedt)
- Terminalia not, or only slightly protruding . . . . . . . . . . . . . . . . . . . . . . . 30
30(29) Abdominal T6 divided, the halves broadly separated
(Fig. 405). ST7 with a pair of glossy black tubercles an-
tero-laterally (Fig. 406) . . . . . . . . . . . . 82. *Parasarcophaga similis* (Meade)
- Terminalia different . . . . . . . . . . . . . . . . . . . . . . . . . . . . . . . . . . . . . . . . . . 31
31(30) White hairs of postgena extending on to anterior part of

gena. Abdominal T6 angular dorsally, with marginal
bristles almost equal in thickness to marginals of T5 (Fig.
362) . . . . . . . . . . . . . . . . . . . . . . . . . 74. *Parasarcophaga albiceps* (Meigen)
–       Gena with black hairs, or with only a few white hairs at
extreme posterior part close to postgena. Terminalia dif-
ferent . . . . . . . . . . . . . . . . . . . . . . . . . . . . . . . . . . . . . . . . . . . . . . . . . . 32
32(31)  Abdominal T8 entire, smooth and shining, and slightly
arching over the cerci (Fig. 393). T6 with a well-diffe-
rentiated row of strong marginal bristles . . . . . . . . . . . . . . . . . . . . . . . . . . .
. . . . . . . . . . . . . . . . . . . . . 80. *Parasarcophaga portschinskyi* Rohdendorf
(Note that 81. *Parasarcophaga pleskei* Rohdendorf may key out here.)
–       Abdominal T8 absent . . . . . . . . . . . . . . . . . . . . . . . . . . . . . . . . . . . . . . . 33
33(32)  Gena with a few white hairs in posterior part close to
postgena . . . . . . . . . . . . . . . . . . . 79. *Parasarcophaga emdeni* Rohdendorf
–       Gena with black hairs only, white hairs entirely restricted
to postgena . . . . . . . . . . . . . . . . . . . . . . . . . . . . . . . . . . . . . . . . . . . . . . . 34
34(33)  Abdominal T6 with strongest marginal bristles in dorsal
position. Membrane below T6 with a median, almost
quadrate sclerotisation (dissection necessary) . . . . . . . . . . . . . . . . . . . . . . . .
. . . . . . . . . . . . . . . . . . . . . . . . . . . . 75. *Parasarcophaga aratrix* (Pandellé)
–       Abdominal T6 with strongest marginal bristles in lateral
position. Membrane below T6 without any sclerotisation . . . . . . . . . . . . 35
35(34)  Abdominal T3 without median marginal bristles. Terminalia black. T6 desclerotised dorsally, the two halves subcontiguous, separated by a long membraneous stripe . . . . . . . . . . . . . . . . .
. . . . . . . . . . . . . . . . . . . . . . . . . . 76. *Parasarcophaga uliginosa* (Kramer)
–       Abdominal T3 most often with, but sometimes without,
a pair of median marginal bristles. Terminalia black or
blackish red. T6 desclerotised dorsally, the two halves
distinctly narrowing medially, separated by a very short
membraneous stripe . . . . . . . . . . . . . . . . . . . . . . . . . . . . . . . . . . . . . . . . . 36
36(35)  Abdominal T8 completely absent (dissection necessary) . . . . . . . . . . . . . . .
. . . . . . . . . . . . . . . . . . . . . . . . . . . 87. *Sarcophaga subvicina* Rohdendorf
–       Abdominal T8 present: divided and often rather weakly
sclerotised, but always with distinct setae . . . . . . . . . . . . . . . . . . . . . . . . . .
84. *Sarcophaga carnaria* (Linnaeus),
85. *Sarcophaga lasiostyla* Macquart,
and 86. *Sarcophaga variegata* (Scopoli)

## Genus *Bellieriomima* Rohdendorf, 1937

*Bellieriomima* Rohdendorf, 1937, Fauna USSR, Dipt., 19 (1): 164.
Type species: *Sarcophaga laciniata* Pandellé, 1896, = *Sarcophaga subulata* Pandellé, 1896.

120

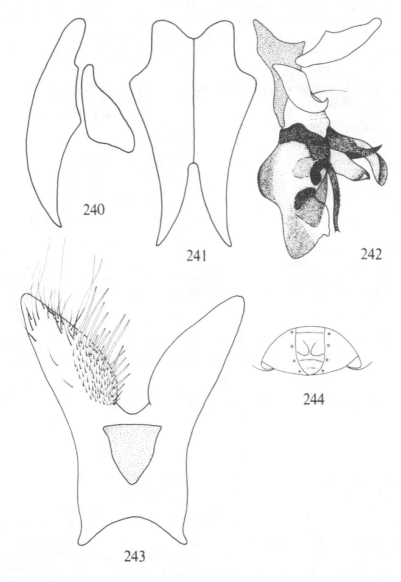

Figs 240-244. *Bellieriomima subulata* (Pandellé). – 240: cerci + surstyli, lateral view; 241: cerci, posterior view; 242: aedeagus; 243: ST5 ♂; 244: terminalia ♀, posterodorsal view.

Medium-sized to large flies. Thorax with 3 post dc.

♂. Ventromedian parts of hind trochanters with long setae. Terminalia black. Protandrial segment pollinose and with marginal bristles. Aedeagus with a well-developed vesica. Median sclerotisation large and laterally compressed, with well-developed median processes. Juxta weakly sclerotised and somewhat hood-like.

About six species in the Palaearctic and Oriental regions.


### 49. *Bellieriomima subulata* (Pandellé, 1896)
Figs 240-244.

*Sarcophaga subulata* Pandellé, 1896, Revue Ent., 15: 194.
*Sarcophaga laciniata* Pandellé, 1896, Revue Ent., 15: 195.

♂. Narrowest part of frons 0.18-0.21×head-width. Thorax with 3 post dc.

Legs: mid femur with a complete row of av and pv bristles, but the latter becoming weaker proximally. The apical pv bristles about as strong as apical av bristles. Hind trochanter with long setae ventromedially. Hind tibia with long hairs on pv and v surfaces. Abdominal T3 without, or with a weak pair of, median marginal bristles. Terminalia black. Protandrial segment densely pollinose and with strong marginal bristles. Aedeagus with large vesical lobes, strong median sclerotisation with well-developed median processes, and a weakly sclerotised juxta (Fig. 242).

♀. Narrowest part of frons 0.28×head-width. Mid femoral organ absent. Terminalia black. T6 desclerotised dorsally, with widely separated halves (Fig. 244) and marginal bristles.

Length ♂♀. 6.0-11.0 mm.

Distribution. Not common; recorded from EJ in Denmark and AK in Norway. Swedish records from Sk., Ång., and Nb. In Finland from the southernmost provinces. – Europe and western USSR.

Biology. Bred from *Lymantria dispar* (Linnaeus) (Lepidoptera: Lymantriidae) (Herting & Simmonds 1976).


## Genus *Discachaeta* Enderlein, 1928

*Discachaeta* Enderlein, 1928, Arch. klassif. phylogen. Ent., 1 (1): 30.
Type species: *Sarcophaga cucullans* Pandellé, 1896.

Small to medium-sized species. Thorax with 3 post dc.

♂. Mid femur without distinct rows of av or pv bristles. Ventromedian parts of hind trochanter with short, stout setae. Abdominal ST5 with numerous short bristles at base of each arm. Terminalia black or red. Protandrial segment somewhat elongated. Cerci more or less excavated dorsally and often with a dorsal angular knob. Aedeagal juxta with distinct basal arm-like processes somewhat similar to the juxtal arms of *Parasarcophaga (Liosarcophaga)*.

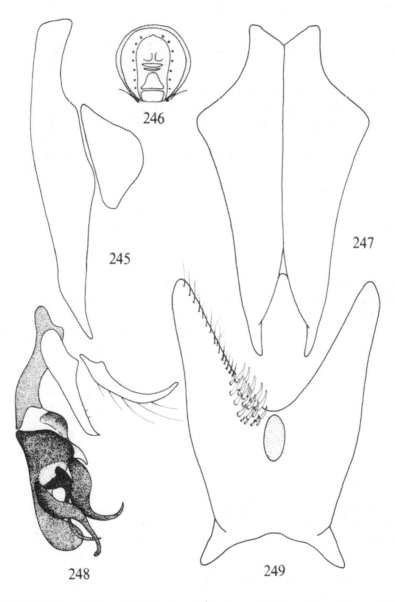

Figs 245-249. *Discachaeta pumila* (Meigen). – 245: cerci + surstyli, lateral view; 246: terminalia ♀, posterior view; 247: cerci, posterior view; 248: aedeagus; 249: ST5 ♂.

♀. Mid femoral organ in distal 0.4-0.5 or absent.

Five Palaearctic species.

## 50. *Discachaeta pumila* (Meigen, 1826)
Figs 245-249.

*Sarcophaga pumila* Meigen, 1826, Syst. Beschr., 5: 24.

♂. Narrowest part of frons 0.18-0.21 × head-width. Lower margin of facial plate somewhat protruding. Thorax with 3 post dc and occasionally with a pair of prst acr. Scutellum without apicals. Legs: mid femur with a row of weak av bristles and without a row of pv. Hind trochanter with short, stout setae ventromedially. Hind femur without a row of av but with a single long preapical av bristle. Hind tibia with long pv hairs. Abdomen only occasionally with a pair of median marginal bristles on T3. Terminalia: protandrial segment dark brown, slightly elongated, with a broad pollinose spot posteriorly, and with marginal bristles. Epandrium dark red or almost black, slightly elongated. Cerci somewhat excavated dorsally. Aedeagal juxta hood-like with long basal arms (Fig. 248). Harpes tapering and diverging distally. Vesica small.

♀. Narrowest part of frons 0.30-0.33 × head-width. Lower margin of facial plate slightly projecting beyond antennal insertion. Mid femoral organ absent. Terminalia black, weakly pollinose. T6 with marginal bristles, undivided but with a small dorsal indentation. Membrane between T6 and cerci without any trace of T7-T8 (Fig. 246).

Length ♂♀. 4.5-6.5 mm.

Distribution. Common in Denmark and southern Fennoscandia. – Widely distributed in Europe east to the Ukraine; North Africa.

Biology. Unknown.

## Genus *Helicophagella* Enderlein, 1928

*Helicophagella* Enderlein, 1928, Arch. klassif. phylogen. Ent., 1 (1): 38.
Type species: *Sarcophaga noverca* Rondani, 1860.

A rather well-defined genus of medium-sized species.

Thorax with 3 post dc.

♂. Each arm of ST5 basally with a tuft of short, stout bristles on inner margin. Aedeagus with harpes directed more or less dorsally.

♀. Terminalia with T6 completely divided.

## 51. *Helicophagella melanura* (Meigen, 1826)
Figs 250-255.

*Sarcophaga melanura* Meigen, 1826, Syst. Beschr., 5: 23.

124

♂. Frons very broad, at narrowest 0.26-0.29×head-width. Thorax with 3 post dc. Legs: mid femur with a complete row of av bristles and an apical row of short pv bristles. Hind trochanter ventro-medially with setae of medium length and a long bristly apical seta. Hind tibia with long hairs on pv and v surfaces. Abdominal T3 without median marginal bristles. Terminalia black. Protandrial segment non-pollinose, or with only traces of pollinosity, and with marginal bristles. Aedeagus with triangular vesical lobes. Harpes reduced to a pair of small forked lobes pointing laterally. Styli rather large and widely separated by the median sclerotisation. Juxta small and weakly sclerotised, almost covering the styli in dried specimens (Fig. 253), but often raised in glycerol preparations (Fig. 252).

♀. Narrowest part of frons 0.35-0.39×head-width. Mid femoral organ absent. Terminalia black. T6 distinctly protruding and densely pollinose; broadly divided and with strong bristles on lateral margins. Membrane between and posterior to the halves

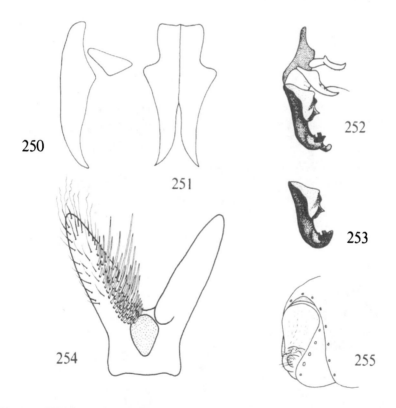

250

251

252

253

254

255

Figs 250-255. *Helicophagella melanura* (Meigen). – 250: cerci + surstyli, lateral view; 251: cerci, posterior view; 252: aedeagus; 253: distiphallus, dried specimen; 254: ST5 ♂; 255: terminalia ♀, posterolateral view.

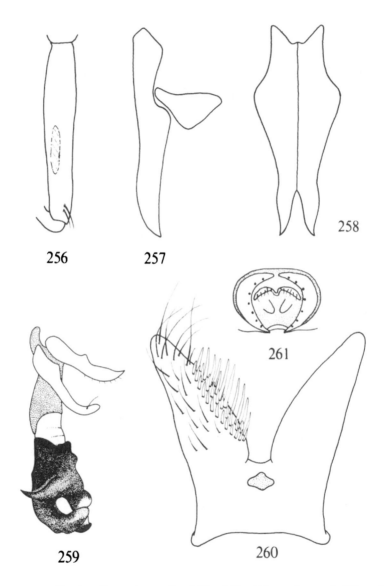

256    257                                    258

261

259                          260

Figs 256-261. *Helicophagella crassimargo* (Pandellé). – 256: mid femoral organ ♀; 257: cerci + surstyli, lateral view; 258: cerci, posterior view; 259: aedeagus; 260: ST5 ♂; 261: terminalia ♀, posterior view.

126

of T6 black and sclerotised, forming a pit-like depression (Fig. 255).
Length ♂♀. 6.0-12.5 mm.

Distribution. Common in Denmark and Fennoscandia north to the Arctic Circle. – Holarctic and Oriental. In the Nearctic region restricted to the temperate parts.

Biology. Breeding in faeces and animal carcasses (Mihályi 1965; Kano et al. 1967). May cause traumatic myiasis in birds and mammals (Séguy 1941; Emden 1954). From Denmark recorded from a hedgehog (Nielsen et al. 1978). Bred from snails (Arion, Helix). Zetterstedt's (1855) record of Sarcophaga striata bred from Oryctes nasicornis (Linnaeus) (Coleoptera: Scarabaeidae) may refer to H. melanura according to Lundbeck (1927).

## 52. *Helicophagella crassimargo* (Pandellé, 1896)
Figs 256-261.

*Sarcophaga crassimargo* Pandellé, 1896, Revue Ent., 15: 195.

♂. Narrowest part of frons 0.18-0.24 × head-width. Thorax with 3 post dc. Legs: mid femur with a complete row of pv and a row of av bristles in proximal 0.5. Hind trochanter ventromedially with setae of medium length and a long bristly apical seta. Hind tibia with a row of long pv hairs. Abdominal T3 with a pair of median marginal bristles, but occasionally without marginals. Terminalia black. Protandrial segment pollinose and with marginal bristles. Cerci in profile narrowest at middle (Fig. 257). Aedeagus with large harpes which, in profile, reach beyond dorsal margin of aedeagus (Fig. 259).
♀. Narrowest part of frons 0.34 × head-width. Mid femoral organ just distal to middle, often somewhat indistinct (Fig. 256). Terminalia black. T6 desclerotised dorsally, lateral margins with bristles which are about 0.3 × as thick as lateral marginals of T5. T8 often present as a pair of sclerites (Fig. 261), but these may be indistinct.
Length ♂♀. 4.5-10.5 mm.

Distribution. Common in Denmark and Fennoscandia except for the northernmost provinces. – Palaearctic region, from the British Isles east to Central Asia.

Biology. A tentative record from the snail *Cernuella virgata* (da Costa) (Keilin 1919).

## 53. *Helicophagella agnata* (Rondani, 1860)
Figs 262-265.

*Sarcophaga agnata* Rondani, 1860, Atti Soc. ital. Sci. nat., 3: 383.

♂. Narrowest part of frons 0.15-0.18 × head-width. Thorax with 3 post dc and often 2-4 prst acr indicated. Legs: mid tibia with 3-4 av bristles in proximal 0.5, with some long pv bristles at middle, and a row of short stout pv bristles in distal 0.5. Hind trochanter with long hairs ventromedially. Hind tibia with rows of long pv and av hairs. Abdominal T3 with a pair of median marginal bristles. Terminalia black. Protandrial segment pollinose and with weak, often hair-like marginal bristles. Aedeagus with a rather compact distiphallus. Harpes curved dorsally, but not reaching dorsal margin. Juxta and styli forming an almost rectangular structure (Fig. 264).

♀. Narrowest part of frons 0.30-0.35 × head width. Mid femoral organ absent. Terminalia black or reddish black. T6 desclerotised dorsally, the halves often distinctly separated. Lateral margins of T6 with bristles which are about 0.7 × as thick as lateral marginals of T5.

Length ♂♀. 6.5-11.0 mm.

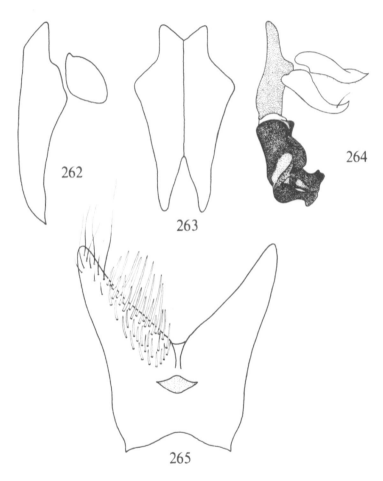

Figs 262-265. *Helicophagella agnata* (Rondani). - 262: cerci + surstyli, lateral view; 263: cerci, posterior view; 264: aedeagus; 265: ST5 ♂.

Distribution. Uncommon in Denmark and southern parts of Norway and Sweden. Not recorded from Finland. – Europe, from the British Isles east to the Ukraine.

Biology. Bred from the snail *Helix aspersa* Müller (Emden 1954).

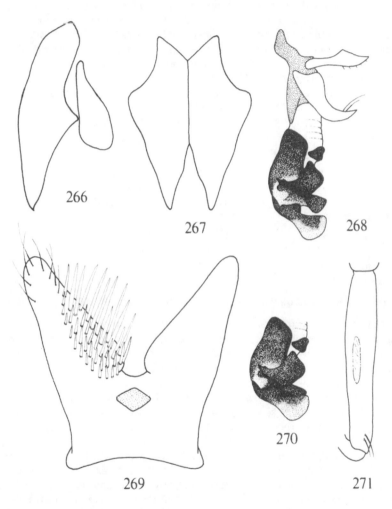

Figs 266-271. *Helicophagella rosellei* (Böttcher). – 266: cerci + surstyli, lateral view; 267: cerci, posterior view; 268: aedeagus; 269: ST5 ♂; 270: distiphallus with juxta covering styli; 271: mid femoral organ ♀.

### 54. *Helicophagella rosellei* (Böttcher, 1912)
Figs 266-271.

*Sarcophaga rosellei* Böttcher, 1912b, Dt. ent. Z., 1912: 714.

♂. Narrowest part of frons 0.19-0.24×head-width. Thorax with 3 post dc. Legs: mid femur with an av row of bristles of variable strength but without distinct pv bristles. Hind trochanter with short setae ventromedially and a bristly apical seta. Hind femur with a row of hair-like av bristles. Abdominal T3 with a pair of median marginal bristles. Terminalia black. Protandrial segment pollinose with marginal bristles. Cerci broad at base and gradually tapering. Aedeagus with harpes just shorter than in *H. crassimargo* and not reaching dorsal margin. Juxta larger and more rounded apically than in *H. crassimargo* (Fig. 268).

♀. Narrowest part of frons 0.30-0.34×head-width. Mid femoral organ almost exactly at middle, of medium size, and often bright red. Terminalia black. T6 desclerotised dorsally, lateral margins with bristles which are only slightly weaker than lateral marginals of T5. T7 present as a pair of easily visible sclerites.
Length ♂♀. 5.5-10.0 mm.

Distribution. Restricted to cool temperate and subarctic provinces in Fennoscandia. Not recorded from Denmark, Finland, or Sweden south of Hrj. – Palaearctic, from the British Isles to the Far East. Alpine in distribution in central and southern Europe, but also recorded from Egypt (Rohdendorf 1934).

Biology. Unknown.

### 55. *Helicophagella noverca* (Rondani, 1860)
Figs 272-275.

*Sarcophaga noverca* Rondani, 1860, Atti Soc. ital. Sci. nat., 3: 386.

♂. Narrowest part of frons 0.17-0.21×head-width. Thorax with 3 post dc. Legs: mid femur with a row of av bristles, at least in proximal 0.6, but without pv bristles. Hind trochanter with short, stout setae ventromedially and several long hairs. Hind femur without distinctly differentiated av or pv bristles except for the subapical av. Hind tibia with long, dense hairs on pv and v surfaces. Abdominal T3 without median marginal bristles. Terminalia black. Protandrial segment with marginal bristles and a large, more or less circular, and densely pollinose spot posteriorly. Aedeagus with a characteristic concave juxta. Harpes curved dorsally, but recurving and uncinnate apically (Fig. 274).

♀. Narrowest part of frons about 0.30-0.31×head width. Mid femoral organ in distal 0.5, large and red. Terminalia black. T6 with a row of strong marginal bristles and a few long hairs. Only a few female specimens have been examined. They are all very similar to females of *H. hirticrus*. The mid femoral organ seems to reach farther distally, and the abdominal T6 possess fewer hairs and has the marginal bristles more regularly placed than in *H. hirticrus*.
Length ♂♀. 10.0-11.5 mm.

Distribution. Not recorded from Denmark or Finland, but a few records from Norway: Ns, and Sweden: Jmt. – Somewhat rare in Europe and western USSR; not in the British Isles.

Biology. Bred from the snail *Helix pomatia* Linnaeus and from horse meat bait (Enderlein 1928; Eberhardt 1955). Séguy (1941) lists *H. noverca* as a coprophile, but Eberhardt (1955) failed to rear the species on faeces. Aradi & Mihályi (1971) caught a single male in a food market.

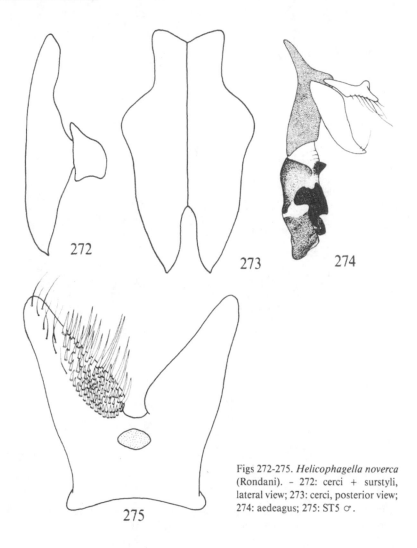

Figs 272-275. *Helicophagella noverca* (Rondani). – 272: cerci + surstyli, lateral view; 273: cerci, posterior view; 274: aedeagus; 275: ST5 ♂.

131

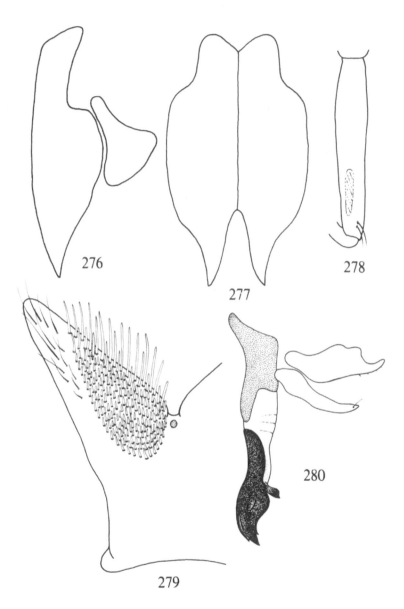

Figs 276-280. *Helicophagella hirticrus* (Pandellé). – 276: cerci + surstyli, lateral view; 277: cerci, posterior view; 278: mid femoral organ ♀; 279: ST5 ♂; 280: aedeagus.

### 56. *Helicophagella hirticrus* (Pandellé, 1896)
Figs 276-280.

*Sarcophaga hirticrus* Pandellé, 1896, Revue Ent., 15: 193.

♂. Narrowest part of frons 0.20-0.23 × head-width. Thorax with 3 post dc and 1-4 prst acr. Legs: fore femur with pv bristles only distinct in distal 0.4-0.5. Mid femur without av or pv bristles. Hind trochanter with long hairs ventromedially. Hind femur without av bristles. Hind tibia with long dense hairs on pv and v surfaces. Abdominal T3 without median marginal bristles. Terminalia black. Protandrial segment pollinose and without marginal bristles. Cerci short, very broad at base, and appearing almost triangular (Fig. 276). Aedeagus with short appendages. Distiphallus long. Vesica present as a pair of small uncinnate lobes (Fig. 280).

♀. Very similar to *H. noverca*. Narrowest part of frons about 0.34 × head-width. Mid femoral organ in distal 0.5 and distinctly red (Fig. 278); variable in size. Terminalia black, T6 with long hairs and strong marginal bristles in a somewhat irregular row. Length ♂♀. 7.5-11.0 mm.

Distribution. Rare in Fennoscandia, only recorded from Sweden: Gtl. and Jmt. Not recorded from Denmark. – Western Palaearctic, from the British Isles to the Ukraine. North Africa.

Biology. Bred from snails of the genus *Helix* (Barfoot 1969; Beaver 1972).


## Genus *Heteronychia* Brauer & Bergenstamm, 1889

*Heteronychia* Brauer & Bergenstamm, 1889, Denkschr. Akad. Wiss. Wien, Kl. math.-naturw., 56 (1): 56.
Type species: *Heteronychia chaetoneura* Brauer & Bergenstamm, 1889, = *Sarcophaga dissimilis* Meigen, 1826.

A large genus of small to medium-sized flies. Thorax with 3 post dc. Wing-vein $R_1$ often setose.

♂. Each arm of abdominal ST5 proximally with a tuft of short, stout bristles at inner margin. Terminalia with protandrial segment and epandrium slender and more or less elongated. Epandrium often red. Cerci flattened dorsally or with a dorsal excavation. Basiphallus long, vesica present as a small scale- or spine-like extension, rarely elongated but never enlarged and bilobed. Membrane proximal to vesica often convex. Juxta often with a pair of flattened, membraneous appendages flanking the apical part of juxta.

♀. Mid femoral organ, if present, situated in apical half.

*Heteronychia* is mainly Palaearctic in distribution, and the greatest diversity seems to be in the Mediterranean area. For a definition of the subgenera, see Rohdendorf (1965).

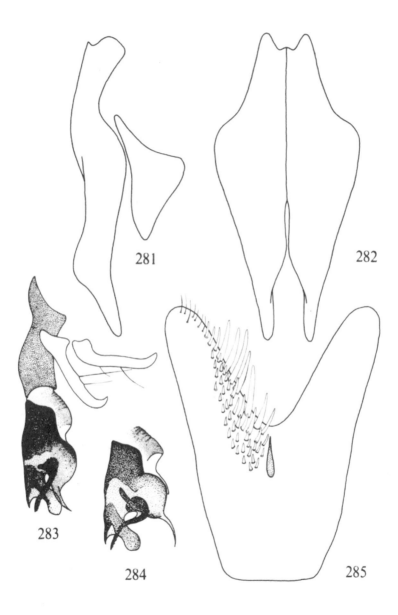

Figs 281-285. *Heteronychia depressifrons* (Zetterstedt). – 281: cerci + surstyli, lateral view; 282: cerci, posterior view; 283: aedeagus; 284:distiphallus with harpes bent outwards; 285: ST5 ♂.

57. *Heteronychia depressifrons* (Zetterstedt, 1845)
Figs 281-285.

*Sarcophaga depressifrons* Zetterstedt, 1845, Dipt. Scand., 4: 1293.
*Pierretia obscurata* Rohdendorf, 1937, Fauna USSR, Dipt., 19 (1): 346.
*Sarcophaga offuscata;* auctt., *nec* Meigen, 1826.

A blackish species, with wing-membrane slightly fumose along the veins.

♂. Narrowest part of frons 0.18-0.22 × head-width. Thorax with 3 post dc and 2-3 prst acr. Scutellar apicals absent, rarely present but then very weak. Wing-vein $R_1$ with 4-7 setae, vein $CuA_1$ occasionally with 1-5 setae. Legs: mid femur with a row of av bristles in about proximal 0.6, and without distinct pv bristles. Hind trochanter ventromedially with stout setae of medium length . Hind tibia with a few somewhat elongated pv hairs, occasionally a complete row. Abdominal T3 with a pair of median marginal bristles, occasionally a more or less distinct pair on T1 + 2. Terminalia black. Protandrial segment shining, with marginal bristles, and with a circular spot of pollinosity posteriorly, seldom totally non-pollinose. Aedeagus with apical part of juxta short, exceeded in length by both styli and juxtal appendages (Fig. 283).

♀. Narrowest part of frons 0.31-0.35 × head-width. Mid femoral organ indistinct. Terminalia black. T6 undivided and with a row of marginal bristles which is interrupted dorsally.

Length ♂♀. 3.5-7.0 mm.

Distribution. Common in Denmark and southern Fennoscandia. - Widely distributed in the Palaearctic and Oriental regions.

Biology. Unknown.


58. *Heteronychia bezziana* (Böttcher, 1913)
Figs 286-288.

*Sarcophaga bezziana* Böttcher, 1913, Dt. ent. Z., 1913: 242.
*Pierretia ostensackeni* Rohdendorf, 1937, Fauna USSR, Dipt., 19 (1): 353.

♂. Narrowest part of frons 0.15 × head-width. Thorax with 3 post dc and 2 prst acr. Scutellar apicals absent, rarely present but then very weak. Wing-vein $R_1$ with 1-3 setae, rarely bare on one or both wings. Legs: mid femur with a row of av bristles in about proximal 0.5 and with a few bristly pv setae. Hind trochanter ventromedially with stout setae of medium length. Hind tibia with a single row of somewhat elongated pv hairs. Abdominal T3 with a pair of median marginal bristles. Terminalia black. Protandrial segment with a spot of pollinosity posteriorly and with marginal bristles. Aedeagus somewhat similar to *H. depressifrons,* but with longer harpes, and without juxtal appendages (Fig. 288).

♀. Like the female *H. depressifrons* but wing-vein $R_1$ with only 1-3 setae.
Length ♂♀. 3.5-6.5 mm.

Distribution. Rare; recorded from Ak in Norway and Vrm. in Sweden. Unknown from Denmark and Finland. – Western Europe; not in the British Isles.

Biology. Unknown.

### 59. *Heteronychia haemorrhoa* (Meigen, 1826)
Figs 289-293.

*Sarcophaga haemorrhoa* Meigen, 1826, Syst. Beschr., 5: 29.

♂. Narrowest part of frons 0.17-0.21×head-width. Thorax with 3 post dc and often some prst acr. Wing-vein $R_1$ setose. Legs: mid femur with a row of av bristles in about proximal 0.6, and without distinct pv bristles although some bristly hairs may be present. Hind trochanter with numerous short setae ventromedially. Hind femur with a row of av hairs which may be more or less bristly. Hind tibia with a row of long pv hairs. Abdominal T3 with a pair of median marginal bristles. Terminalia: protandrial segment brownish black and shining, with a large, almost circular pollinose spot posteriorly, and with marginal bristles. Epandrium red. Cerci in profile with a distinct dorsal swelling. Aedeagus with apical part of juxta long, distinctly longer than juxtal appendages (Fig. 293).

♀. Narrowest part of frons 0.26-0.31×head-width. Mid femoral organ in distal

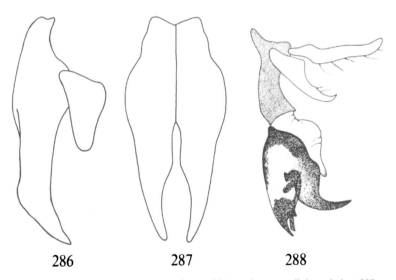

286          287          288

Figs 286-288. *Heteronychia bezziana* (Böttcher). – 286: cerci + surstyli, lateral view; 287: cerci, posterior view; 288: aedeagus.

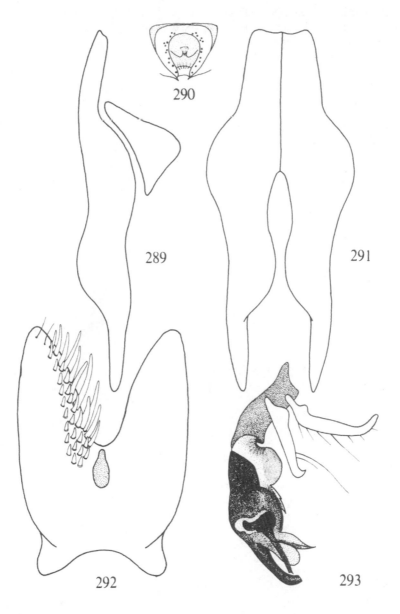

Figs 289-293. *Heteronychia haemorrhoa* (Meigen). – 289: cerci + surstyli, lateral view; 290: terminalia ♀, posterior view; 291: cerci, posterior view; 292: ST5 ♂; 293: aedeagus.

137

half, sometimes absent. Terminalia red. T6 undivided, evenly curved, and with a single row of marginal bristles which is interrupted dorsally (Fig. 290).

Length ♂♀. 5.0-9.5 mm.

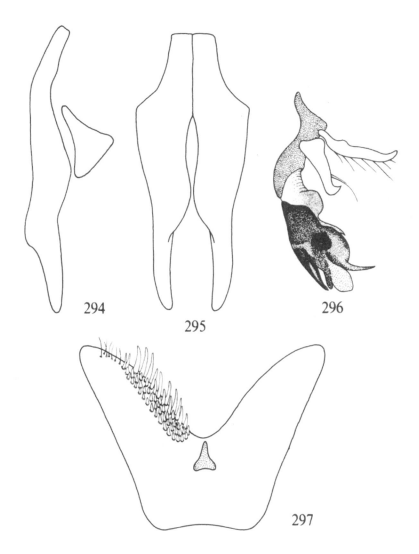

Figs 294-297. *Heteronychia boettcheriana* (Rohdendorf). – 294: cerci + surstyli, lateral view; 295: cerci, posterior view; 296: aedeagus; 297: ST5 ♂.

138

Distribution. Common in Denmark and southern Fennoscandia. – Western Palae-arctic, from the British Isles east to the Ukraine.

Biology. Bred from the snail *Cepaea hortensis* (Müller) (Mik 1890; Schmitz 1917).

Note. The row of av bristles on hind femur is often weak, and this character is used in the keys of Rohdendorf (1937) and Emden (1954). However, it is variable and speci-mens with a row of strong av bristles are not infrequent.

### 60. *Heteronychia boettcheriana* (Rohdendorf, 1937)
Figs 294-297.

*Pierretia boettcheriana* Rohdendorf, 1937, Fauna USSR, Dipt., 19 (1): 345.

♂. Narrowest part of frons 0.15-0.20 × head-width. Thorax, wings, abdomen, and legs as in *H. haemorrhoa*, but av bristles of hind femur never hair-like. Terminalia: pro-tandrial segment with a large pollinose spot posteriorly which is about 2 × as broad as long. Aedeagus with apical part of juxta of medium length, just shorter than juxtal ap-pendages (Fig. 296).
♀. Indistinguishable from females of *H. haemorrhoa*.
Length ♂♀. 6.0-11.0 mm.

Distribution. Common in southern Sweden and southern Norway. Only a single male from Denmark: NEZ (BMNH). Not recorded from Finland. – Europe east to the Ukraine and the Caucasus. Not in the British Isles.

Biology. Unknown.

### 61. *Heteronychia vagans* (Meigen, 1826)
Figs 298-302.

*Sarcophaga vagans* Meigen, 1826, Syst. Beschr., 5: 26.
*Sarcophaga frenata* Pandellé, 1896, Revue Ent., 15: 182.

♂. Narrowest part of frons 0.19-0.21 × head-width. Thorax with 3 post dc. Wing-vein $R_1$ setose or bare. Legs: mid femur with a row of av bristles in proximal 0.6, and an api-cal row of short stout pv bristles. Hind trochanter with short stout setae ventromedial-ly. Hind femur with a complete row of strong av bristles. Hind tibia with a row of long pv hairs. Abdominal T3 with a pair of median marginal bristles. Terminalia with red epandrium. Protandrial segment brownish black and shining, with an oval pollinose spot posteriorly (longest axis in median plane), and a row of marginal bristles. Cerci with a small indentation at apex (Fig. 298). Aedeagus with membrane proximal to vesi-ca greatly swollen (Fig. 301). Juxta with very short appendages.
♀. Narrowest part of frons 0.30-0.33 × head-width. Mid femoral organ as in *H. hae-morrhoa*. Terminalia red. T6 slightly desclerotised and angular on dorsal median line

139

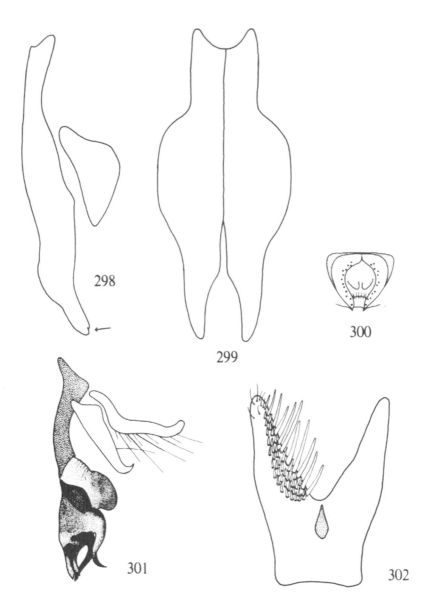

Figs 298-302. *Heteronychia vagans* (Meigen). – 298: cerci + surstyli, lateral view; 299: cerci, posterior view; 300: terminalia ♀, posterior view; 301: aedeagus; 302: ST5 ♂.

(Fig. 300), laterally with strong marginal bristles and bristly hairs almost forming a double marginal row.

Length ♂♀. 6.0-10.0 mm.

Distribution. Very common and widely distributed in Denmark and Fennoscandia, reaching the northernmost provinces in Norway and Sweden. – Widely distributed in the Palaearctic region from the British Isles to Japan.

Biology. Bred from snails of the family Succineidae (Verves 1976b). Schiner (1862) reports specimens of *H. vagans* bred from human flesh, but the identification must be regarded as erroneous.


## 62. *Heteronychia vicina* (Macquart, 1835)
Figs 303-307.

*Sarcophaga vicina* Macquart, 1835, Hist. nat. Ins., Dipt., 2: 225.
*Sarcophaga ruficauda* Zetterstedt, 1838, Insecta Lapp.: 650.
*Sarcophaga ebrachiata* Pandellé, 1896, Revue Ent., 15: 182.

♂. Narrowest part of frons 0.17-0.24 × head-width. Thorax with 3 post dc. Wing-vein $R_1$ bare or with a single seta. Legs: mid femur with a row of av bristles in proximal 0.6, and an apical row of long pv bristles. Hind trochanter with short stout setae ventromedially. Hind femur without av bristles except for the subapical. Hind tibia with long pv hairs. Abdominal T3 without median marginal bristles (but a single female with 1 median marginal bristle has been seen). Terminalia with red epandrium. Protandrial segment brownish black and shining, with an oval pollinose spot posteriorly (longest axis in median plane), and marginal bristles. Cerci in profile straight and gradually tapering, dorsal face slightly excavated. Aedeagus with a long distiphallus. Harpes large and oriented perpendicular to longitudinal axis of aedeagus.

♀. Narrowest part of frons 0.31-0.34 × head-width. Mid femoral organ in distal half (Fig. 303), but distinct and often bright red. Terminalia red. T6 very weakly desclerotised dorsally; somewhat intermediate between *H. haemorrhoa* and *H. vagans*.

Length ♂♀. 7.0-10.0 mm.

Distribution. Not recorded from Denmark. Common in Norway and northern Sweden. Recorded from southern Finland. – Boreo-montane distribution. Recorded from the Pyrenees, the Alps, the Caucasus, and other mountainous regions in central and southern Europe; British Isles.

Biology. Unknown.


## 63. *Heteronychia proxima* (Rondani, 1860)
Figs 308-311.

*Sarcophaga proxima* Rondani, 1860, Atti Soc. ital. Sci. nat., 3: 392.

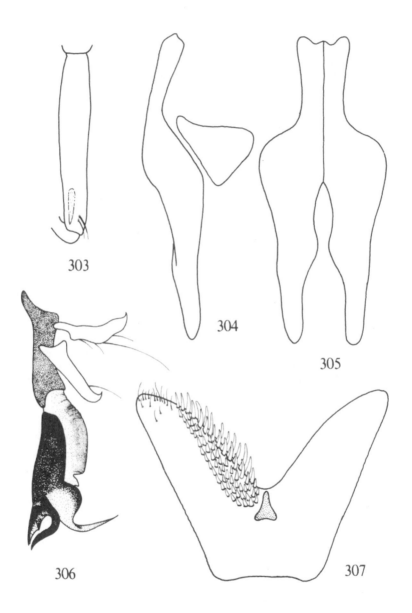

Figs 303-307. *Heteronychia vicina* (Macquart). – 303: mid femoral organ ♀; 304: cerci + sursty-li, lateral view; 305: cerci, posterior view; 306: aedeagus; 307: ST5 ♂.

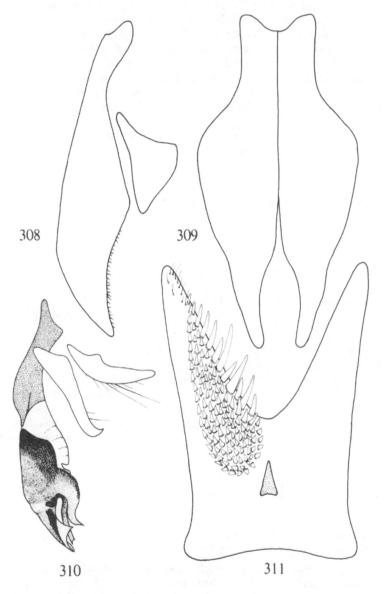

308     309

310     311

Figs 308-311. *Heteronychia proxima* (Rondani). – 308: cerci + surstyli, lateral view; 309: cerci, posterior view; 310: aedeagus; 311: ST5 ♂.

♂. Narrowest part of frons 0.17-0.18×head-width. Thorax with 3 post dc. Some prst acr may be differentiated among the hairs. Wing-vein R$_1$ bare. Legs: mid femur with a row of av bristles in proximal 0.6. Hind trochanter with short setae ventromedially. Hind femur with a row of both av and pv bristles. Hind tibia without or with a few long pv hairs. Abdominal T3 most often without median marginal bristles, but specimens from Finland possess well-developed median marginals on T3. Terminalia with red epandrium. Protandrial segment shining black with a broad, more or less circular pollinose spot posteriorly, and marginal bristles. Cerci with a very slight excavation dorsally. Aedeagus with a slightly swollen membrane below vesica. Harpes rather short (Fig. 310).

♀. Very similar to *H. vicina,* but somewhat variable in the presence of mid femoral organ and median marginal bristles on T3.

Length ♂♀. 8.0-10.0 mm.

Distribution. Not recorded from Denmark or Norway. A few records from Sweden: Öl., and Finland: N. – Palaearctic, from Europe and western part of USSR east to China. Not in the British Isles.

Biology. Unknown.

## Genus *Pierretia* Robineau-Desvoidy, 1863

*Pierretia* Robineau-Desvoidy, 1863, Hist. nat. Dipt. Paris, 2: 422.
Type species: *Pierretia praecox* Robineau-Desvoidy, 1863, = *Sarcophaga nigriventris* Meigen, 1826.

Small to medium-sized flies. Thorax with 3 post dc and often some prst acr. Abdominal T3 with a pair of strong median marginal bristles.

♂. Ventromedian part of hind trochanter with long setae. Abdominal ST5 with bristly hairs along inner margin of each arm. Terminalia black. Cerci rather broad in profile, with an apical hook. Aedeagus with lateral parts of distiphallus desclerotised. Median sclerotisation blade-like. Juxta often well-sclerotised and deeply cleft into two closely approximated prongs, but less sclerotised and undivided in *P. sexpunctata.*

♀. Mid femoral organ absent or present in distal 0.3.

### 64. *Pierretia sexpunctata* (Fabricius, 1794)
Figs 312-315.

*Musca sexpunctata* Fabricius, 1794, Ent. syst., 4: 300.
*Sarcophaga clathrata* Meigen, 1826, Syst. Beschr., 5: 25.

Easily recognisable in both sexes by the black-haired occiput and postgena.

♂. Narrowest part of frons 0.17-0.20×head-width. Parafacial setae moderately developed, not long and bristly. Postgenal hairs black, with only a few white hairs around occipital foramen. Thorax with 3 post dc and 1-3 prst acr. Prescutellar acr always well-developed. Scutellar apicals always present. Wing-vein $R_1$ occasionally with 1-4 hairs. Legs: mid femur with a complete row of av and pv bristles, the apical pv bristles shorter and stronger than apical av bristles. Hind tibia with long pv hairs in varying numbers. Abdomen with a pair of strong median marginal bristles on T3. Terminalia black. Pro-

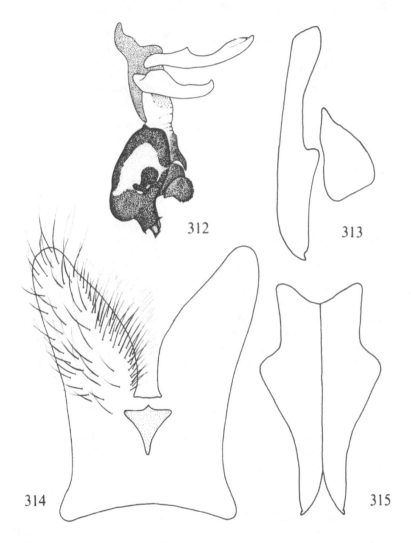

Figs 312-315. *Pierretia sexpunctata* (Fabricius). – 312: aedeagus; 313: cerci + surstyli, lateral view; 314: ST5 ♂; 315: cerci, posterior view.

tandrial segment with distinct pollinosity on posterior 0.5, this occasionally rather sparse. Hairs on posterior margin slightly stronger, but never developed as distinct marginal bristles. Cerci with dorsal margin evenly convex. Aedeagus with well-developed vesica which is slightly recurved and serrated along anterior margin. Harpes indistinct and hidden behind the vesical lobes. Juxta sclerotised and evenly rounded, not tapering or divided apically, but with apical margin extended laterally into pointed processes.

♀. Narrowest part of frons 0.28-0.31 × head-width. Terminalia with T6 divided dorsally.

Length ♂♀. 5.5-8.5 mm.

Distribution. Common in Denmark and Fennoscandia, reaching the northernmost provinces in Sweden and Finland. – Widely distributed in the Palaearctic region.

Biology. Bred from egg-cocoons of spiders: *Araneus cornutus* Clerk and *Clubiona* sp. (Mik 1890; Lundbeck 1927; Grunin 1964), and from various acridid grasshoppers (see Séguy 1941 for references). Kano *et al.* (1967) failed to breed the species from various kinds of vertebrate carrion.

## 65. *Pierretia nemoralis* (Kramer, 1908)
Figs 316-319.

*Sarcophaga nemoralis* Kramer, 1908, Ent. Wbl., 25: 152.

♂. Narrowest part of frons 0.22-0.24 × head-width. Parafacial setae long and bristly. Postgenal hairs white. Thorax with 3 post dc and sometimes 1-2 prst acr. Prescutellar acr always well-developed. Scutellar apicals always present. Legs: mid femur with a complete row of av and pv bristles, the apical pv bristles shorter and stronger than apical av bristles. Hind tibia with long hairs, varying in numbers from a single pv row to rather dense pv and v hairs. Abdomen with a pair of strong median marginal bristles on T3. Terminalia black. Protandrial segment with distinct pollinosity on posterior 0.5, this occasionally rather sparse. Hairs on posterior margin often slightly stronger, but never developed as distinct marginal bristles. Cerci with dorsal margin evenly convex. Aedeagus with vesica three-lobed and with slightly serrated margins (Fig. 318). Harpes weakly sclerotised and bifurcate apically. Juxta weakly sclerotised and with an apical cleft.

♀. Narrowest part of frons about 0.30-0.34 × head-width. Abdominal T6 not or only slightly divided dorsally. Easily distinguished from all other *Pierretia* by the distinct red or yellowish mid femoral organ, which is situated in distal 0.3.

Length ♂♀. 6.5-10.5 mm.

Distribution. Not recorded from Denmark. A single female from Norway: AK and a few records from Sweden: Upl., Jmt., and Finland: Ta, Lk. I have not been able to confirm the record from Sweden: Hrj. in Ringdahl (1952). – Central and northern Europe, USSR to the Far East. Povolný & Šustek (1983) record *P. nemoralis* as a montane species.

Biology. Unknown.

Note. I have not been able to recover any females caught in copula, and the assignment to *P. nemoralis* of the only three females seen (Norway: AK in Coll. Rognes; Sweden: Upl. in Coll. Bergström; Finland: Ta in ZMUC) is somewhat tentative.

### 66. *Pierretia villeneuvei* (Böttcher, 1912)
Figs 320-323.

*Sarcophaga villeneuvei* Böttcher, 1912a, Dt. ent. Z., 1912: 347.

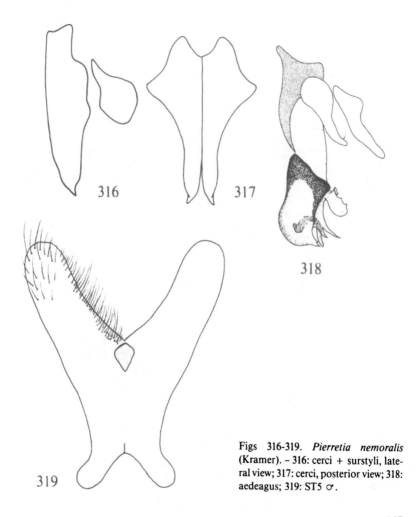

Figs 316-319. *Pierretia nemoralis* (Kramer). – 316: cerci + surstyli, lateral view; 317: cerci, posterior view; 318: aedeagus; 319: ST5 ♂.

147

White postgenal hairs not extending on to posterior part of gena. Arista with longest hairs 1.5-3.0× as long as second aristomere.

♂. Narrowest part of frons 0.27-0.32× head-width. Parafacial setae long and bristly. Thorax with 3 post dc and 1-2 prst acr. Prescutellar acr indistinct or moderately developed. Scutellum with 2 pairs of lateral bristles, 1 pair of discals, but no apicals. Legs: mid femur with a complete row of av and pv bristles, the apical pv bristles about as long and strong as apical av bristles. Hind tibia with a few long pv hairs. Abdomen with a pair of strong median marginal bristles on T3. Terminalia black and shining. Protandrial segment non-pollinose and without marginal bristles. Aedeagus with

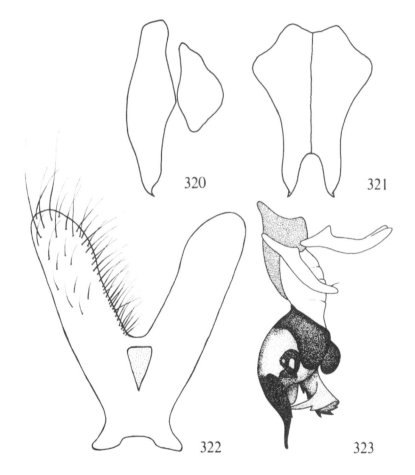

Figs 320-323. *Pierretia villeneuvei* (Böttcher). – 320: cerci + surstyli, lateral view; 321: cerci, posterior view; 322: ST5 ♂; 323: aedeagus.

148

small vesica and large, recurved harpes which are serrated along median margin (Fig. 323). Juxta long and deeply cleft.

♀. Narrowest part of frons about 0.35-0.36 × head-width. Abdominal T6 distinctly divided dorsally. Very similar to females of *P. socrus,* but the aristal hairs apparently shorter.

Length ♂♀. 4.5-6.5 mm.

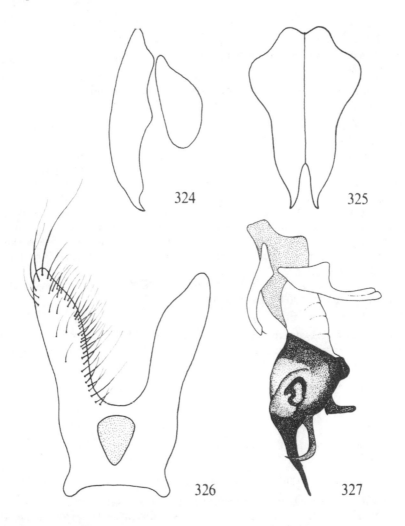

Figs 324-327. *Pierretia socrus* (Rondani). – 324: cerci + surstyli, lateral view; 325: cerci, posterior view; 326: ST5 ♂; 327: aedeagus.

149

Distribution. Rare but probably widespread in Denmark and Fennoscandia. From NEZ in Denmark. Not recorded from Norway. A few records from Sweden and Finland. - Europe and western USSR, east to the Altai Mountains and China.

Biology. Unknown.

### 67. *Pierretia socrus* (Rondani, 1860)
Figs 324-327.

*Sarcophaga socrus* Rondani, 1860, Atti Soc. ital. Sci. nat., 3: 387.
*Sarcophaga rostrata* Pandellé, 1896, Revue Ent., 15: 203.

White postgenal hairs not extending on to posterior part of gena. Arista with longest hairs 3.0-4.5× as long as second aristomere.

♂. Narrowest part of frons 0.23-0.28 × head-width. Parafacial setae long and bristly. Thorax with 3 post dc and 1-3 prst acr. Prsc acr indistinct or well-developed. Scutellar apical bristles always present. Legs: mid femur with a complete row of av and pv bristles, the apical pv bristles about as long and strong as apical av bristles. Hind tibia without or with a few long pv hairs, but these never with wavy tips. Abdomen with a pair of strong median marginal bristles on T3. Terminalia black and shining. Protandrial segment non-pollinose, and without marginal bristles. Aedeagus with vesica long and bent at a right angle. Harpes long and recurved. Juxta elongated, deeply cleft, and gradually tapering (Fig. 327).

♀. Narrowest part of frons 0.31-0.36× head-width. Very similar to females of *P. villeneuvei,* but the aristal hairs apparently longer.

Length ♂♀. 5.0-8.0 mm.

Distribution. Recorded from Finland: Al, Sa. Verves (1986) lists *P. socrus* from Denmark, but this needs confirmation. Not known from Norway or Sweden. - Europe except for the British Isles, western USSR.

Biology. Unknown.

### 68. *Pierretia nigriventris* (Meigen, 1826)
Figs 328-331.

*Sarcophaga nigriventris* Meigen, 1826, Syst. Beschr., 5: 27.

Gena with some white hairs on posterior part close to postgena.

♂. Narrowest part of frons 0.27-0.33 × head-width. Parafacial setae long and bristly. Arista with longest hairs 1.5-3.0× as long as second aristomere. Thorax with 3 post dc and 1-3 prst acr. Prsc acr indistinct or moderately developed. Scutellar apicals always present. Legs: mid femur with a complete row of av and pv bristles, the apical pv bristles about as long and strong as apical av bristles. Hind tibia without or with a few long pv hairs, but these never with wavy tips. Abdomen with a pair of strong median marginal bristles on T3. Terminalia black and shining. Protandrial segment non-pollinose, and without marginal bristles. Aedeagus with well-developed vesica. Harpes straight and lanceolate; as long as vesica. Dorsal sclerotisation of distiphallus extending beyond base of juxta (Fig. 331). Apical part of juxta cleft.

♀. Narrowest part of frons 0.35-0.39× head-width. Indistinguishable from females of *P. soror.*

Length ♂♀. 4.5-8.0 mm.

Distribution. Common in Denmark, but not recorded from Fennoscandia. – Widely distributed in Europe, North Africa, and USSR to the Far East.

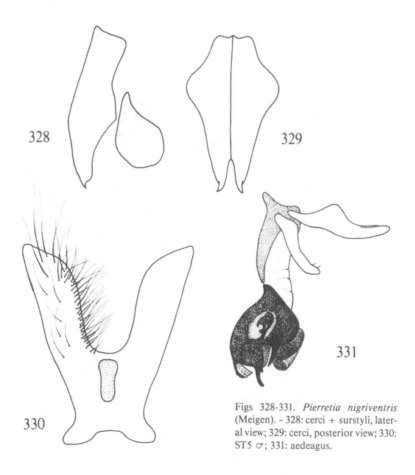

Figs 328-331. *Pierretia nigriventris* (Meigen). – 328: cerci + surstyli, lateral view; 329: cerci, posterior view; 330: ST5 ♂; 331: aedeagus.

151

Biology. Parasitising, preying, or scavenging upon snails and various arthropods. Bred from snails: *Cernuella, Eobania, Helicella, Helix,* and *Monacha* (Böttcher 1913; Bowell 1917; Keilin 1919; Séguy 1953; Miles 1968; Barfoot 1969; Beaver 1972; Cameron & Disney 1975). Bred from acridid grasshoppers: *Schistocerca gregaria* (Forskål) (Séguy 1932), from beetles: *Carabus, Necrophorus,* and possibly *Blaps* (Emden 1950),

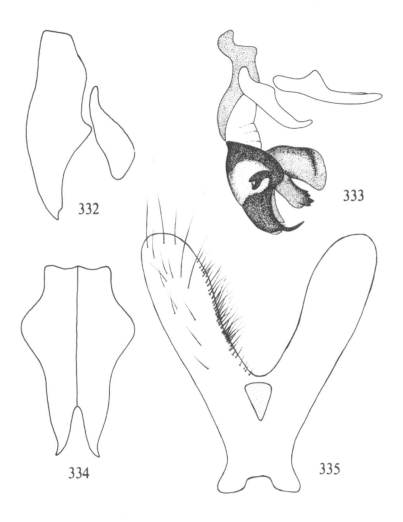

Figs 332-335. *Pierretia soror* (Rondani). - 332: cerci + surstyli, lateral view; 333: aedeagus; 334: cerci, posterior view; 335: ST5 ♂.

and from the honeybee *Apis mellifera* Linnaeus (Guilhon 1945; Séguy 1965). A female recorded in Smith (1957) as bred from the bumblebee *Bombus terrestris* (Linnaeus), and identified as either *P. nigriventris* or *P. villeneuvei,* belongs to the present species.

Larvae mature within the maternal uterus and are nourished by secretions from the accessory glands. The female fly larviposits second-instar larvae directly on to a suitable host (Séguy 1965).

### 69. *Pierretia soror* (Rondani, 1860)
Figs 332-335.

*Sarcophaga soror* Rondani, 1860, Atti Soc. ital. Sci. nat., 3: 386.

Gena with some white hairs on posterior part close to postgena.

♂. Narrowest part of frons 0.25-0.28 × head width. Parafacial setae long and bristly. Thorax with 3 post dc and 2-3 prst acr. Prsc acr indistinct or moderately developed. Scutellar apicals always present. Legs: mid femur with a complete row of av and pv bristles, the apical pv bristles about as long and strong as apical av bristles. Hind tibia with long hairs on pv and v surfaces, the longest hairs with wavy tips. Abdomen with a pair of strong median marginal bristles on T3. Terminalia black and shining. Protandrial segment non-pollinose and without marginal bristles. Aedeagus with well-developed vesica. Harpes as long as vesica, serrated along anterior margin. Juxta deeply cleft, slender apical part almost perpendicular to longitudinal axis of aedeagus (Fig. 333).

♀. Indistinguishable from females of *P. nigriventris.*
Length ♂♀: 4.5-9.0 mm.

Distribution. Not rare in Denmark and southern Norway and Sweden. Not recorded from Finland. – Europe and western USSR. Recorded from Ireland (Richards 1960) but not from Great Britain.

Biology. Bred from the snail *Helix aspersa* (Müller) (Séguy 1921).

## Genus *Sarcotachinella* Townsend, 1892

*Sarcotachinella* Townsend, 1892, Trans. Am. ent. Soc., 19: 110.

Type species: *Sarcotachinella intermedia* Townsend, 1892, = *Sarcophaga sinuata* Meigen, 1826.

Thorax with 3 post dc. Mid femur with golden hairs. Abdominal T3 with a pair of median marginal bristles.

♂. Ventromedian parts of hind trochanters with short, stout setae. Each arm of abdominal ST5 with strong bristles on inner margin and long hairs apically. Terminalia black. Cerci subapically with some short, thorn-like setae. Aedeagus with a large, three-lobed vesica. Harpes flat and oriented transversely.

A single Holarctic species.

153

70. *Sarcotachinella sinuata* (Meigen, 1826)
   Figs 336-340.

*Sarcophaga sinuata* Meigen, 1826, Syst. Beschr., 5: 22.

Easily recognised in both sexes by the distinct golden hairs on mid femur.

♂. Narrowest part of frons 0.22-0.27 × head-width. Thorax with 3 post dc. Legs: mid femur with a complete row of av and pv bristles; with distinct golden hairs in the distal 0.3 of anterior surface, and often some golden hairs on posterior surface just distal to middle. Hind tibia with a few long pv hairs.

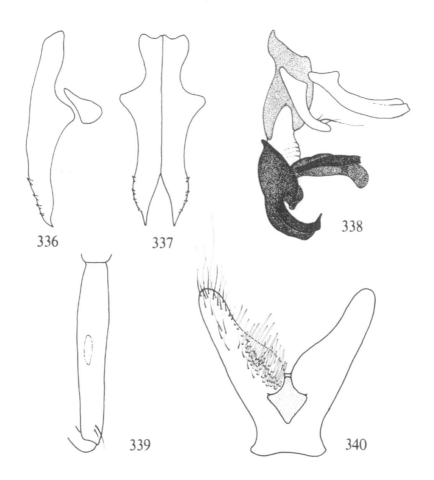

Figs 336-340. *Sarcotachinella sinuata* (Meigen). – 336: cerci + surstyli, lateral view; 337: cerci, posterior view; 338: aedeagus; 339: mid femoral organ ♀; 340: ST5 ♂.

154

♀. Narrowest part of frons 0.33-0.34×head-width. Mid femoral organ small, situated at middle just ventral to the golden hairs (Fig. 339). Terminalia protruding; T6 grey pollinose, desclerotised dorsally, and with strong bristles on lateral margins. Anus situated in a small, often sclerotised, pit-like depression.

Length ♂♀. 4.5-8.0 mm.

Distribution. Common in Denmark and southern and central Fennoscandia, reaching as far north as Lu.Lpm. in Sweden. – Holarctic.

Biology. Attracted to faeces and dead organic matter (Rohdendorf 1959; Aradi & Mihályi 1971). Bred from acridid grasshoppers (*Melanoplus* spp.) in the Nearctic region (Smith 1958). Aldrich (1916: postscript) mentions specimens of *S. sinuata* attacking flying grasshoppers for larviposition.

A female from Finland (ZMH) was bred from *Nonagria typhae* (Thunberg) (Lepidoptera: Noctuidae).

## Genus *Thyrsocnema* Enderlein, 1928

*Thyrsocnema* Enderlein, 1928, Arch. klassif. phylogen. Ent., 1 (1): 42.

Type species: *Musca striata* Fabricius, 1794, sensu Enderlein; misidentification, = *Sarcophaga incisilobata* Pandellé, 1896.

Thorax with 3 post dc. Abdominal T3 without median marginal bristles.

♂. Hind trochanters with short setae on ventromedian parts. Terminalia black. Protandrial segment pollinose, without marginal bristles. Cerci slightly swollen apically, with a well-developed hook. Aedeagus with vesica divided into median and lateral lobes, the latter distinctly spinose. Median processes greatly elongated. Juxta well-developed, tapering into a blunt or pointed tip and with a pair of processes arising very close to each-other.

♀. Mid femoral organ medium-sized and situated at middle. Abdominal T6 divided dorsally, with widely-spaced marginal bristles and long marginal hairs.

Four species of *Thyrsocnema* in the Palaearctic region and the mountainous areas bordering the Palaearctic and Oriental regions.

Most authors have employed a rather different concept of *Thyrsocnema*. Rohdendorf (1963a) listed *Thyrsocnema* as a junior synonym of *Pierretia* Robineau-Desvoidy, whilst Zumpt (1972) included species generally assigned to *Liosarcophaga* Enderlein (a subgenus of *Parasarcophaga* Johnston & Tiegs). Rohdendorf (1965) raised *Thyrsocnema* sensu stricto to generic rank, but Verves (1986) includes *Pseudothyrsocnema* Rohdendorf and *Nudicerca* Rohdendorf as subgenera within *Thyrsocnema*.

71. *Thyrsocnema incisilobata* (Pandellé, 1896)
    Figs 341-344.

*Sarcophaga incisilobata* Pandellé, 1896, Revue Ent., 15: 197.

155

♂. Narrowest part of frons 0.22-0.24×head-width. Thorax with 3 post dc. Legs: mid femur with a complete row of av bristles which may be somewhat indistinct apically in small specimens, and a complete row of pv. Hind tibia with long hairs on pv and v surfaces. Abdomen without median marginal bristles on T3, although a very weak pair may be indicated. Each arm of ST5 with a distinct bristly swelling on inner mar-

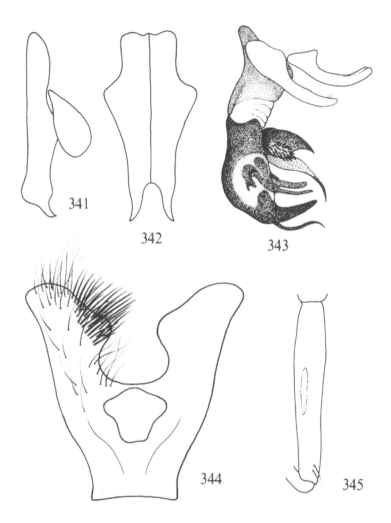

Figs 341-344. *Thyrsocnema incisilobata* (Pandellé). – 341: cerci + surstyli, lateral view; 342: cerci, posterior view; 343: aedeagus; 344: ST5 ♂.
Fig. 345. *Thyrsocnema* sp. mid femoral organ ♀.

gin. Terminalia black. Protandrial segment covered with pollinosity and without marginal bristles. Cerci swollen apically and with a well-developed hook. Aedeagus with apical part of juxta long, median processes smoothly curved, and spinose lobes of vesica shorter than median lobes (Fig. 343).

♀. Narrowest part of frons 0.32-0.34 × head-width. Mid femoral organ at middle and of medium size (Fig. 345). Scutellum often with 3 pairs of lateral bristles. Terminalia black to reddish. T6 desclerotised dorsally, with marginal bristles, and, posterior to these, long hairs. Indistinguishable from *T. kentejana*.

Length ♂♀. 6.0-11.0 mm.

Distribution. Very common in Denmark and recorded from all provinces. Common in southern Fennoscandia, reaching Vrm. in Sweden and HO in Norway. In Finland only from the southernmost provinces. – Widely distributed in the Palaearctic region;

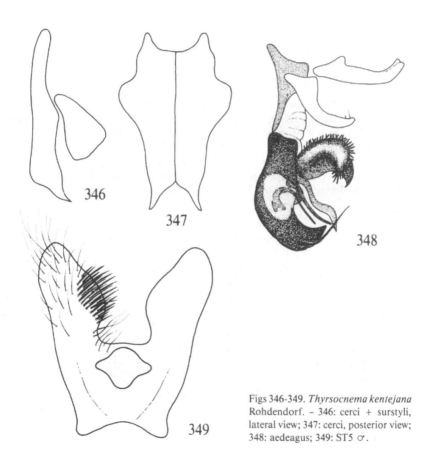

Figs 346-349. *Thyrsocnema kentejana* Rohdendorf. – 346: cerci + surstyli, lateral view; 347: cerci, posterior view; 348: aedeagus; 349: ST5 ♂.

157

from the British Isles east to Central Asia, and from southern Scandinavia south to North Africa.

Biology. Larvae living in faeces. Reported as bred from a snail (Keilin 1919) and as a parasitoid on the acridid grasshopper *Stauronotus maroccanus* (Thunberg) (Séguy 1941).

## 72. *Thyrsocnema kentejana* Rohdendorf, 1937
Figs 346-349.

*Thyrsocnema kentejana* Rohdendorf, 1937, Fauna USSR, Dipt., 19 (1): 174.
*Thyrsocnema kentejana lapponica* Tiensuu, 1939, Annls ent. fenn., 5: 265.

♂. Narrowest part of frons 0.21-0.23 × head-width. Thorax, legs, and abdomen as in *T. incisilobata*. Terminalia black. Protandrial segment covered with pollinosity and without marginal bristles. Cerci swollen apically and with a well-developed hook, which is slightly longer than in *T. incisilobata*. Aedeagus with apical part of juxta of medium length, median processes with a subapical bend, and spinose lobes of vesica as long as or longer than median lobes (Fig. 348).
♀. Indistinguishable from *T. incisilobata*.
Length ♂♀. 5.5-10.0 mm.

Distribution. Not found in Denmark. In Fennoscandia mainly from the northern provinces, but reaching southwards to Hrj. in Sweden and TE in Norway. The latter record is interesting as it indicates sympatry with *T. incisilobata,* but *T. kentejana* seems to be restricted to the low arctic/alpine zone while *T. incisilobata* is found exclusively in the lowlands in the northern part of its range. – Boreo-montane in distribution: northern Scandinavia, France, Austria, Switzerland, Romania, USSR (Siberia, Sikhote-Alin Mountains), India (Kashmir), Mongolia, North China.

Biology. Unknown.

## Genus *Bercaea* Robineau-Desvoidy, 1863

*Bercaea* Robineau-Desvoidy, 1863, Hist. nat. Dipt. Paris, 2: 549.
Type species: *Musca haemorrhoidalis* Fallén, 1816, sensu Meigen, 1826; misidentification, = *Sarcophaga cruentata* Meigen, 1826.

Large flies. The white hairs of postgena and occiput extend at least on to posterior half of gena. Thorax with 5-6 post dc, but the anterior 3-4 of these are often rather weak. Abdominal T3 without median marginal bristles.
♂. Mid femur with a complete row of av and an apical row of short stout pv bristles. Ventromedian parts of hind trochanter with numerous short stout setae. Terminalia red or yellowish, but protandrial segment often somewhat blackish. Protandrial segment pollinose and with distinct marginal bristles. Cerci more or less flattened in transverse plane and distinctly separated distally. Aedeagus with a long, well-sclerotised distiphallus. Vesical lobes flattened in horizontal plane.

♀. Terminalia bright red to orange. T6 deeply divided dorsally. Five Afrotropical species one of which is cosmopolitan.

### 73. *Bercaea cruentata* (Meigen, 1826)
Figs 350-356.

*Sarcophaga cruentata* Meigen, 1826, Syst. Beschr., 5: 28.
*Sarcophaga haemorrhoidalis;* auctt. *nec* Fallén, 1817.

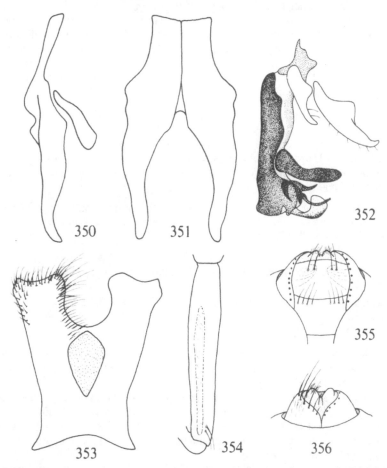

Figs 350-356. *Bercaea cruentata* (Meigen). – 350: cerci + surstyli, lateral view; 351: cerci, posterior view; 352: aedeagus; 353: ST5 ♂; 354: mid femoral organ ♀; 355: terminalia ♀, ventral view; 356: terminalia ♀, dorsal view.

Gena with white hairs on at least posterior 0.5.

♂. Narrowest part of frons 0.24-0.28 × head-width. Thorax with 5-6 post dc, but the anterior 3-4 often weak and hair-like. Prsc acr absent or hair-like. Legs: mid femur with a complete row of av bristles and an apical row of short stout pv bristles. Hind tibia with long dense hairs on pv and v surfaces. Abdomen without median marginal bristles on T3. Terminalia red to yellowish, with protandrial segment somewhat blackish. Protandrial segment pollinose on posterior 0.5 and with few but distinct marginal bristles. Cerci broadly separated distally and somewhat flattened in transverse plane; in profile with a distinct notch (Fig. 350). Aedeagus with a long well-sclerotised distiphallus. Vesica strong and bilobed (Fig. 352).

♀. Narrowest part of frons 0.30-0.35 × head-width. Scutellum with or without a weak additional lateral bristle. Mid femoral organ large (Fig. 354). Terminalia bright red to orange. T6 with a single, well-defined row of marginal bristles; divided dorsally into a deep notch (Fig. 356). ST6-ST7 broad and fully visible. ST6 with a marginal row of 6-10 bristles, ST7 with a triangular elevation in middle and 2-4 bristles (Fig. 355).
Length ♂♀. 7.0-15.0 mm.

Distribution. Recorded from Denmark: NEZ (Copenhagen), where it seems to be entirely restricted to urban areas, probably taking advantage of the locally favourable microclimate for hibernation. Unknown from Fennoscandia. – Cosmopolitan; recorded from all zoogeographical regions except the Australian, but most common in the tropics and the warmer parts of subtropical regions. Probably spread from the Afrotropical region.

Biology. Breeding in various kinds of decomposing organic matter, but mainly in faeces. Larval development from newly-deposited larvae to adults may take from 8 to 16 days (Knipling 1936; Zumpt 1965).

## Genus *Parasarcophaga* Johnston & Tiegs, 1921

*Parasarcophaga* Johnston & Tiegs, 1921, Proc. R. Soc. Qd, 33: 86.
    Type species: *Sarcophaga omega* Johnston & Tiegs, 1921, = *Sarcophaga sericea* Walker, 1852.

A very species-rich genus of generally large flies. The genus appears somewhat heterogeneous, and a subdivision into several mostly well-defined subgenera is generally used (Rohdendorf 1937, 1965; Verves 1986).

No unambiguous definition of *Parasarcophaga* has been proposed, and the genus may be paraphyletic with regard to several Oriental genera. I prefer to retain the genus in a wide sense instead of uncritically raising the subgenera to generic rank, as this would actually contribute little to the elucidation of the highly problematic infratribal phylogeny of the Sarcophagini.

For a definition of the subgenera see Rohdendorf (1937, 1965). Note that Zumpt's (1972) concept of *Thyrsocnema* includes species generally assigned to *Parasarcophaga* *(Liosarcophaga)*.

## 74. *Parasarcophaga albiceps* (Meigen, 1826)
Figs 357-362.

*Sarcophaga albiceps* Meigen, 1826, Syst. Beschr., 5: 22.

♂. Narrowest part of frons 0.20-0.26 × head-width. White hairs on postgena extending

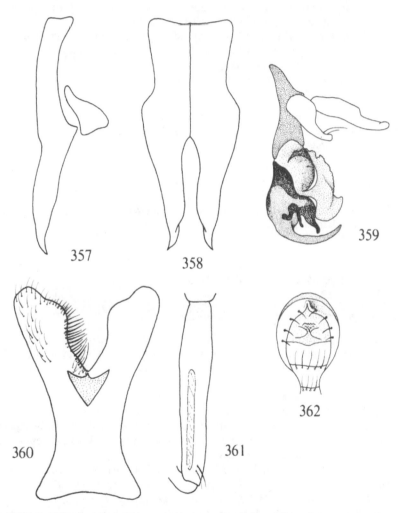

Figs 357-362. *Parasarcophaga albiceps* (Meigen). – 357: cerci + surstyli, lateral view; 358: cerci, posterior view; 359: aedeagus; 360: ST5 ♂; 361: mid femoral organ ♀; 362: terminalia ♀, posterior view.

on to anterior half of gena. Thorax with 4-5 post dc, but the anterior 2-3 often rather weak. Legs: mid femur with a complete row of av bristles and an apical row of short pv. Hind trochanter with short stout setae ventromedially. Hind tibia with long hairs on pv and v surfaces. Abdomen without median marginal bristles on T3, but occasionally with a pair of decumbent bristly setae. Terminalia black. Protandrial segment densely pollinose and without marginal bristles, but some bristly setae may be present on posterior margin. Aedeagus with bilobed juxta and an unpaired vesica which divides proximally into two recurving processes (Fig. 359).

♀. Narrowest part of frons 0.27-0.30 × head-width. Fore femoral organ indistinct. Mid femoral organ large (Fig. 361). Terminalia black to reddish black. Abdominal T6 with bristles about as strong as marginals of T5 and with a distinct dorsal fold (Fig. 362). ST6-ST7 with a row of marginal bristles.

Length ♂♀. 9.0-13.0 mm.

Distribution. Not recorded from Denmark. A few records in the southern parts of Norway and Sweden. Rather common in Finland north to Kb and Om. – Widely distributed in the Palaearctic and Oriental regions.

Biology. Breeding in carcasses, faeces, and garbage (Mihályi 1965; Kano *et al.* 1967). Recorded as a parasitoid/predator of lepidopterous pupae: *Aporia, Dendrolimus, Lymantria;* and beetle larvae: *Oryctes, Melolontha, Polyphylla, Saperda* (Kleine 1910; Thompson 1951; Herting & Simmonds 1976). Senior-White *et al.* (1940) report a case of dermal myiasis in a bull.

75. *Parasarcophaga aratrix* (Pandellé, 1896)
Figs 363-369.

*Sarcophaga aratrix* Pandellé, 1896, Revue Ent., 15: 191.

♂. Narrowest part of frons 0.18-0.24 × head-width. Thorax with 4-5 post dc. Legs: mid femur with a complete row of av bristles and an apical row of short pv bristles. Hind trochanter with short stout setae ventromedially. Hind tibia with long pv and v hairs. Abdomen usually without median marginal bristles on T3, but occasionally a weak pair may be present. Median part of ST5 strongly raised, in profile forming a distinct tooth which projects between the ventral margins of T6 (Fig. 368). Terminalia black. Protandrial segment pollinose and without marginal bristles. Aedeagus with large styli which are supported by the harpes. Juxta well-developed, lateral arms represented by a short tooth at juxtal base (Fig. 365).

♀. Narrowest part of frons 0.30-0.33 × head-width. Fore femoral organ absent. Mid femoral organ large (Fig. 366). Terminalia black. T6 slightly desclerotised dorsally, with 1-2 rows of strong marginal bristles and some long hairs. Strongest marginal bristles on T6 situated dorsally. Membrane below T6 with a median, almost quadrate sclerotisation. ST7 with a few hairs in postero-lateral position (Fig. 369).

Length ♂♀. 9.0-13.5 mm.

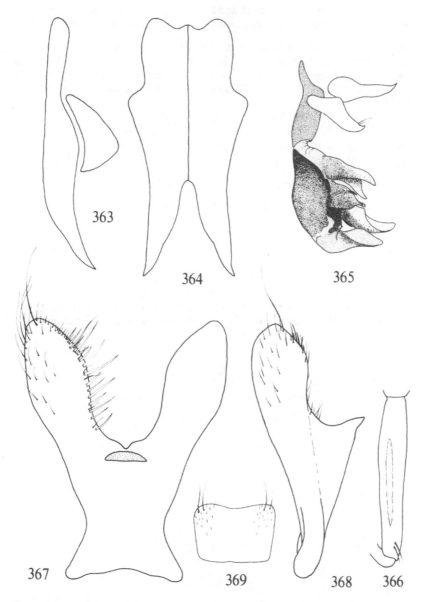

Figs 363-369. *Parasarcophaga aratrix* (Pandellé). – 363: cerci + surstyli, lateral view; 364: cerci, posterior view; 365: aedeagus; 366: mid femoral organ ♀; 367: ST5 ♂, ventral view; 368: ST5 ♂, lateral view; 369: ST7 ♀.

Distribution. Common in Denmark and Fennoscandia north to the Arctic Circle. – Palaearctic region, from the British Isles east to Kamtchatka. Nearctic Region.

Biology. Breeding in decomposing meat (Eberhardt 1955; Kuusela & Hanski 1982). Bred from *Lymantria monacha* (Linnaeus) (Lepidoptea: Lymantriidae). Bred from *Prionus coriarius* (Linnaeus) (Coleoptera: Cerambycidae) in Denmark and Finland (Lundbeck 1927; Saalas 1943). Blackith & Blackith (1984) give an interesting account on larval behaviour.

### 76. *Parasarcophaga uliginosa* (Kramer, 1908)
Figs 370-372.

*Sarcophaga uliginosa* Kramer, 1908, Ent. Wbl., 25: 152.

♂. Narrowest part of frons 0.15 × head-width. Thorax with 4 post dc and 1-4 prst acr, the latter more or less distinct. Legs: mid femur with a complete row of av and an apical row of short, stout pv bristles. Hind trochanter with short stout setae ventromedially. Hind tibia with long pv hairs. Abdomen without median marginal bristles on T3. Terminalia black. Protandrial segment pollinose and without marginal bristles. Cerci distinctly hollowed out dorsally. Aedeagus with a large bilobed juxta. Lateral arms of juxta represented by a short tooth at juxtal base (Fig. 372).

♀. Very similar to females of *Sarcophaga* spp., but without median marginal bristles on abdominal T3. Narrowest part of frons 0.36 × head width. Terminalia black. T6

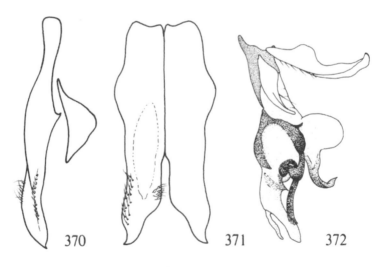

Figs 370-372. *Parasarcophaga uliginosa* (Kramer). – 370: cerci + surstyli, lateral view; 371: cerci, posterior view; 372: aedeagus.

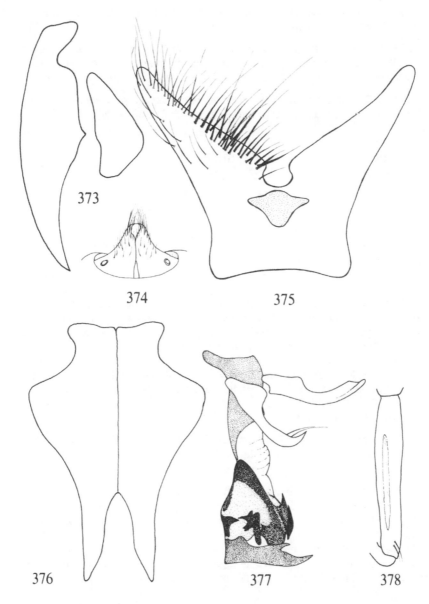

Figs 373-378. *Parasarcophaga caerulescens* (Zetterstedt). – 373: cerci + surstyli, lateral view; 374: terminalia ♀, dorsal view; 375: ST5 ♂; 376: cerci, posterior view; 377: aedeagus; 378: mid femoral organ ♀.

membraneous along a narrow dorsal stripe. T8 absent (as in *Sarcophaga subvicina*). Length ♂♀. 11.0-17.0 mm.

Distribution. Rare, two males from Denmark: NEZ. Not known from Fennoscandia. – Widely distributed in the Palaearctic region, from western Europe east to China and Japan. Oriental region (Korea), and northern parts of Nearctic region.

Biology. The larvae are predatory on lepidopterous pupae: *Dendrolimus, Euproctis, Euxoa, Lymantria, Orgyia, Porthetria* (Herting & Simmonds 1976).


77. *Parasarcophaga caerulescens* (Zetterstedt, 1838)
    Figs 373-378.

*Sarcophaga caerulescens* Zetterstedt, 1838, Insecta Lapp., 650.
*Sarcophaga scoparia* Pandellé, 1896, Revue Ent., 15: 189.

Blackish species.

♂. Narrowest part of frons 0.21-0.25 × head-width. Thorax with 4 post dc. Legs: mid femur with a complete row of av and an apical row of pv bristles. Hind trochanter ventromedially with long and moderately long hairs; without short stout setae. Hind tibia with long pv hairs. Abdomen with long and dense hairs; no median marginal bristles on T3. Terminalia black. Protandrial segment densely pollinose and with marginal bristles. Epandrium flattened dorsally or with a slight concavity. Aedeagus with an undivided vesica. Styli short. Juxta extended into two apically forked arms, which form an almost complete circle in apical view (Fig. 377).

♀. Narrowest part of frons 0.29-0.34 × head-width. Scutellum with an additional lateral bristle. Mid femoral organ large (Fig. 378). Terminalia distinctly protruding and densely pollinose. T6 divided dorsally (Fig. 374) and with a dense row of marginal bristles. ST6 fully visible.

Length ♂♀. 10.0-16.0 mm.

Distribution. Common in Denmark and Fennoscandia, reaching as far north as T.Lpm. in Sweden and LkW in Finland. – Widely distributed in the Palaearctic region. Nearctic (Alaska) and Oriental (Korea) regions.

Biology. Necrophagous, breeding in animal carcasses (Kano *et al.* 1967; Sýchevskaya 1970) and reared on decaying beef and liver (Mihályi 1965; Hanski & Kuusela 1980). The larvae may live as predators of lepidopterous pupae (Rohdendorf & Verves 1978).

Oviposition takes place mainly in late summer with only a single generation per year. Adults emerge in early to mid summer. The species prefers shaded semi-woodland habitats (Hanski & Kuusela 1980; Kuusela & Hanski 1982).


78. *Parasarcophaga jacobsoni* Rohdendorf, 1937
    Figs 379-382.

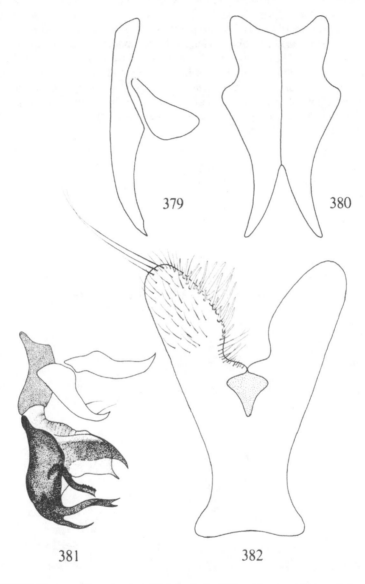

379

380

381

382

Figs 379-382. *Parasarcophaga jacobsoni* Rohdendorf. – 379: cerci + surstyli, lateral view; 380: cerci, posterior view; 381: aedeagus; 382: ST5 ♂.

*Parasarcophaga jacobsoni* Rohdendorf, 1937, Fauna USSR, Dipt., 19 (1): 220.

♂. Narrowest part of frons 0.24-0.26 × head-width. White hairs on postgena extending on to posterior half of gena. Thorax with 4 post dc. Legs: mid femur with a complete row of av and an apical row of short stout pv bristles. Hind trochanter with short stout setae ventromedially. Hind tibia with long pv hairs. Abdomen without median marginal bristles on T3. ST5 distinctly keeled. Terminalia with protandrial segment blackish brown and epandrium reddish or occasionally almost blackish brown. Protandrial segment without marginal bristles. Aedeagus with well-sclerotised vesica which is as long as, or slightly longer than, harpes. Apical part of juxta long (Fig. 381).

♀. Terminalia red to dark red. T6 undivided and with strong marginal bristles. Length. ♂♀. 7.0-14.0 mm.

Distribution. Rare, a single male from Denmark: NEJ (Læsø) in ZMUC. No records from Fennoscandia. – Widely distributed in the Palaearctic region, from western Europe to the Far East, Mongolia, and North China. Not in the British Isles. Oriental region (Korea).

Biology. Necrophagous (Rohdendorf & Verves 1978). Adults may be attracted to food markets (Mihályi 1966).

### 79. *Parasarcophaga emdeni* Rohdendorf, 1969
Figs 383-386.

*Parasarcophaga emdeni* Rohdendorf, 1969, Ént. Obozr., 48: 946.

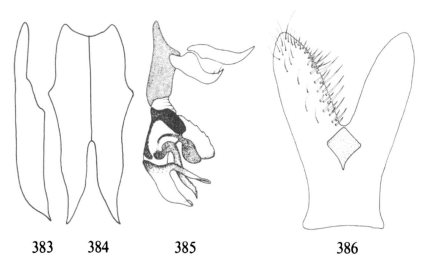

383        384              385                          386

Figs 383-386. *Parasarcophaga emdeni* Rohdendorf. - 383: cerci, lateral view; 384: cerci, posterior view; 385: aedeagus; 386: ST5 ♂.

Gena with a few white hairs on posterior part close to postgena, rarely with black hairs only.

♂. Narrowest part of frons 0.18-0.24 × head-width. Thorax with 4 post dc. Legs: mid femur with a complete row of av and an apical row of short pv bristles. Hind trochanter with short, stout setae. Hind tibia with a row of long pv hairs. Abdomen without median marginal bristles on T3. ST5 slightly keeled. Terminalia black. Protandrial segment pollinose and without marginal bristles. Aedeagus with slender unforked juxtal arms (Fig. 385). Apical part of juxta membraneous and deeply bilobed.

♀. Narrowest part of frons about 0.32 × head-width. Mid femoral organ large. Terminalia black. T6 desclerotised dorsally, with 1-2 rows of strong marginal bristles and some long hairs. Strongest marginal bristles on T6 situated dorsally. Membrane below T6 with a median, almost quadrate sclerotisation.

Length ♂♀. 8.0-13.5 mm.

Distribution. Rare in Denmark and southern Fennoscandia. – Palaearctic region, east to the Altai Mountains and northern China. Not in the British Isles.

Biology. Adults attracted to faeces and collected in food markets (Rohdendorf 1959; Aradi & Mihályi 1971). Bred from the snail *Helicella obvia* (Menke) (Verves & Kuz'movich 1979).

### 80. *Parasarcophaga portschinskyi* Rohdendorf, 1937
Figs 387-393.

*Parasarcophaga portschinskyi* Rohdendorf, 1937, Fauna USSR, Dipt., 19 (1): 226.

♂. Narrowest part of frons 0.21-0.25 × head-width. Thorax with 4 post dc. Legs: mid femur with a complete row of av and an apical row of pv bristles. Hind trochanter with short stout setae ventromedially. Hind tibia with long pv hairs. Abdomen without median marginal bristles on T3. ST5 distinctly keeled, but profile much less angular than in *P. pleskei* (Fig. 391). Terminalia black. Protandrial segment thinly pollinose and without marginal bristles. Gonopod apically with a strongly curved hook (Fig. 389). Aedeagus with a slender, partly membraneous vesica which is distinctly shorter than harpes. Styli terminating well before tip of juxtal arms. Apical part of juxta short.

♀. Narrowest part of frons 0.34-0.35 × head-width. Mid femoral organ large. Terminalia red to dark blackish red. T6 entire but with a weak median longitudinal fold. Marginal bristles on T6 strong and close together, T6 otherwise with very short hairs. T8 visible in undissected specimens, totally bare, and often shining; somewhat arching over the cerci (Fig. 393).

Length ♂♀. 7.5-15.0 mm.

Distribution. Uncommon but widely distributed in Denmark and Fennoscandia, reaching north of the Arctic Circle in Sweden. – Widely distributed in the temperate and subarctic parts of the Palaearctic region. Not in the British Isles.

Biology. Necrophagous, and an occasional predator of lepidopterous larvae (Roh-

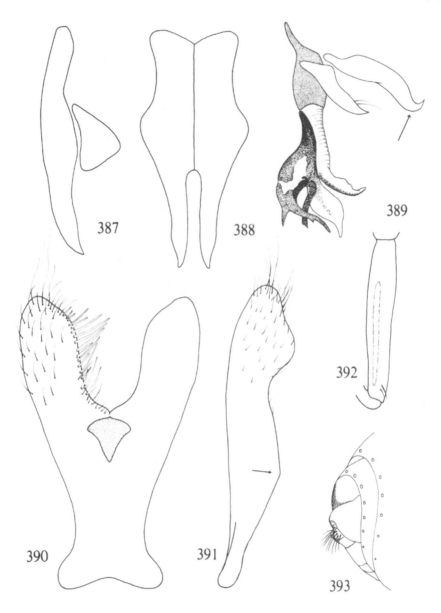

Figs 387-393. *Parasarcophaga portschinskyi* Rohdendorf. – 387: cerci + surstyli, lateral view; 388: cerci, posterior view; 389: aedeagus; 390: ST5 ♂, ventral view; 391: ST5 ♂, lateral view; 392: mid femoral organ ♀; 393: terminalia ♀, posterolateral view.

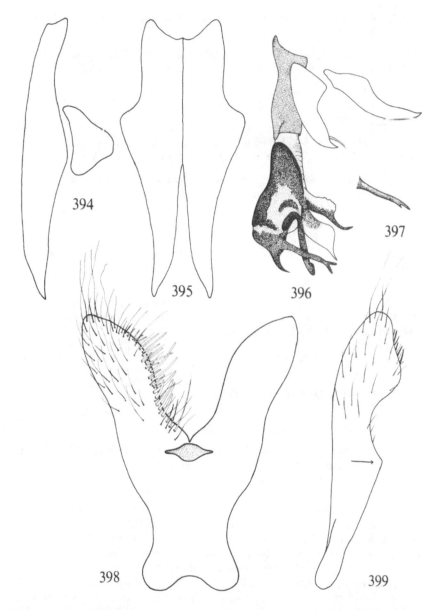

Figs 394-399. *Parasarcophaga pleskei* Rohdendorf. – 394: cerci + surstyli, lateral view; 395: cerci, posterior view; 396: aedeagus; 397: juxtal arm with short dorsal prong; 398: ST5 ♂, ventral view; 399: ST5 ♂, lateral view.

dendorf & Verves 1978; Zhang 1982). A male from southern Finland (ZMH) was bred from a grass snake.

Note. The apical bifurcation of the juxtal arms is somewhat variable: the dorsal prong is often short or indistinct, but occasionally the prongs are subequal.

### 81. *Parasarcophaga pleskei* Rohdendorf, 1937
Figs 394-399.

*Parasarcophaga pleskei* Rohdendorf, 1937, Fauna USSR, Dipt., 19 (1): 230.
*Sarcophaga tuberosa verticina* Ringdahl, 1945, Opusc. ent., 10: 35.

♂. Very similar to *P. portschinskyi*, but with the following differences: abdominal ST5 with a distinctly convex base but not sharply keeled, and with a distinctly angular profile (Fig. 399). Gonopods gradually tapering. Aedeagus with long styli which terminate at apex of juxtal arms. Apical part of juxta long. Apical bifurcation of juxtal arms variable (Figs 396, 397).
♀. The female of *P. pleskei* is still unknown to me. It will probably key out to *P. portschinskyi*.
Length ♂. 8.0-13.0 mm.

Distribution. Somewhat rare; a few records from Norway and northern Sweden. Not recorded from Denmark or Finland. – Widely distributed in temperate and subarctic parts of the Holarctic region.

Biology. Unknown.

### 82. *Parasarcophaga similis* (Meade, 1876)
Figs 400-406.

*Sarcophaga similis* Meade, 1876, Entomologist's mon. Mag., 12: 268.

♂. Narrowest part of frons 0.18-0.22× head-width. Thorax with 4 post dc. Legs: mid femur with a complete row of av and an apical row of short pv bristles. Hind trochanter with short setae and numerous long hairs ventromedially, occasionally with only long hairs. Hind tibia with a row of slightly elongated pv hairs. Abdomen without median marginal bristles on T3. Each arm of ST5 proximally with a tuft of short stout bristles (Fig. 403). Terminalia black. Protandrial segment pollinose and without marginal bristles. Aedeagus with long tapering vesical lobes. Juxtal arms slender and unforked, inclined towards vesica but distinctly recurving subapically. Juxta terminating in a sclerotised pointed tip (Fig. 402).
♀. Narrowest part of frons about 0.30× head-width. Fore femoral organ weakly developed. Mid femoral organ large (Fig. 404). Terminalia black. T6 divided and with widely separated halves (Fig. 405); each half with marginal bristles and some long hairs. ST7 with a pair of glossy black tubercles (Fig. 406).
Length ♂♀. 9.0-12.5 mm.

Distribution. Not recorded from Denmark. Rather common in the southern parts of Norway and Sweden. Very common in southern Finland. – Widely distributed in the Palaearctic and Oriental regions.

Biology. Reared from decaying beef (Mihályi 1965). In Japan it is a common synantropic species, breeding in excrement and carcasses and causing intestinal myiasis (Kano *et al.* 1967). In Finland bred from larvae of *Lacanobia oleracea* (Linnaeus)

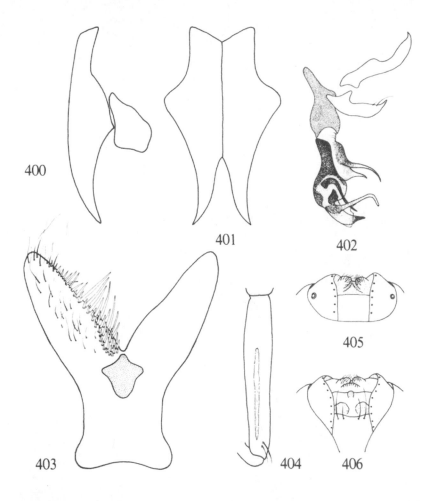

Figs 400-406. *Parasarcophaga similis* (Meade). – 400: cerci + surstyli, lateral view; 401: cerci, posterior view; 402: aedeagus; 403: ST5 ♂; 404: mid femoral organ ♀; 405: terminalia ♀, dorsal view; 406: terminalia ♀, ventral view.

173

(Lepidoptera: Noctuidae), and from dead nestlings of the House Martin (Tiensuu 1939).

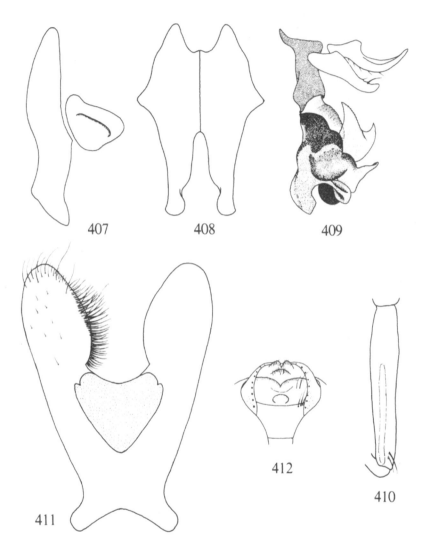

Figs 407-412. *Parasarcophaga argyrostoma* (Robineau-Desvoidy). – 407: cerci + surstyli, lateral view; 408: cerci, posterior view; 409: aedeagus; 410: mid femoral organ ♀; 411: ST5 ♂; 412: terminalia ♀, ventral view.

83. *Parasarcophaga argyrostoma* (Robineau-Desvoidy, 1830)
   Figs 407-412.

*Myophora argyrostoma* Robineau-Desvoidy, 1830, Essai Myod.: 340.
*Sarcophaga barbata* Thomson, 1869, K. svenska Fregatten Eugenies Resa, 2 (1): 533.
*Sarcophaga falculata* Pandellé, 1896, Revue Ent., 15: 185.

♂. Narrowest part of frons 0.24-0.26×head-width. Gena with white hairs except for some black hairs on upper anterior part. Legs: fore femur often with pv bristles hair-like. Mid femur with a complete row of av and an apical row of short stout pv bristles. Hind trochanter with short stout setae ventromedially. Hind tibia with long pv and v hairs. Abdominal T3 without median marginal bristles. ST5 with a large window (Fig. 411). Terminalia with red epandrium. Protandrial segment black and pollinose, with marginal bristles which may be almost hair-like. Cerci angular at dorsal subapical margin, and surstyli distinctly thickened along basal margin (Fig. 407). Aedeagus with a large distiphallus. Styli thickened and juxtal arms flattened apically (Fig. 409).
   ♀. Narrowest part of frons 0.33×head-width. Fore femoral organ distinct and with a few cross-striations. Mid femoral organ large. Terminalia red or yellow. T6 undivided and with marginal bristly hairs. ST7 with a median tubercle near anterior margin, and 1-2 pairs of bristles on posterior margin (Fig. 412).
   Length ♂♀. 9.0-16.0 mm.

Distribution. In Denmark only from NEZ (Copenhagen). No Fennoscandian records. – A cosmopolitan species, recorded from all zoogeographical regions except the Australian.

Biology. Breeding in decaying meat. The larvae have been recorded from cases of myiasis in humans (Sacca 1945; James 1947; Burgess 1966), and of secondary myiasis in sheep (Baranov & Jezic 1928).
   Larvae predatory on beetles, lepidopterous larvae, and egg pods of acridid grasshoppers. Bred from snails in Denmark (Lundbeck 1927). The larval development from newly-deposited larva to adult takes about 16 days under favourable conditions (Hafez 1940).

# Genus *Sarcophaga* Meigen, 1826

*Sarcophaga* Meigen, 1826, Syst. Beschr., 5: 14.
   Type species: *Musca carnaria* Linnaeus, 1758.

Medium-sized to large flies. Thorax with 4 post dc. Abdominal T3 usually with a pair of median marginal bristles.
   ♂. Mid femur with an apical row of short pv bristles, and hind trochanter with short stout setae on ventromedian part. Abdominal ST5 with long hairs along inner margin of each arm. Terminalia black. Aedeagus with distinct ventral plates, vesica sclerotised and bladder-like, styli well-developed and close together, and juxta often weakly sclerotised.

♀. Scutellum often with 3 pairs of lateral scutellar bristles. Fore femoral organ more or less distinct. Mid femoral organ large. Terminalia dark red to black.

Species of *Sarcophaga* seem to be obligate predators of earthworms (Oligochaeta: Lumbricidae), although they may be reared in decaying meat under laboratory conditions. The larvae are deposited in or near the entrances of earthworm burrows, and they probably seek out their prey actively. The point of larval attack is often at the clitellum.

Some authors, e.g. Roback (1954), Downes (1965), and Zumpt (1972), have used a broad concept of the genus *Sarcophaga*. In the present restricted sense, the genus contains about 20 species from the Palaearctic region, and a single species from South Africa, known from the holotype only (?mislabelled). The subdivisions of Lehrer (1973b) are not accepted.

The nomenclature of *Sarcophaga* is under revision (Richet *in prep.*), and the names used in the present paper are those which I find the most correct. Future modifications may be expected.


## 84. *Sarcophaga carnaria* (Linnaeus, 1758)
Figs 415, 417, 418, 421; pl. 2: 4.

*Musca carnaria* Linnaeus, 1758, Syst. Nat. ed. 10, 1: 596.
*Sarcophaga carnaria* var. *schulzi* Müller, 1922, Arch. Naturgesch., 88A (2): 91.
*Sarcophaga subvicina vulgaris* Rohdendorf, 1937, Fauna USSR, Dipt., 19 (1): 287.
*Sarcophaga dolosa* Lehrer, 1967, Zool. Anz., 178 (3-4): 215.

♂. Narrowest part of frons 0.20-0.23 × head-width. Thorax with 4 post dc. A few prst acr may be differentiated. Legs: mid femur with a complete row of av bristles and an apical row of short pv bristles. Hind trochanter with numerous short stout setae ventromedially. Hind tibia with long hairs on p, pv, and v surfaces. Abdomen with a pair of median marginal bristles on T3, but occasionally these may be rather weak or absent. Terminalia black. Protandrial segment non-pollinose or with faint traces of pollinosity, and with marginal bristles. Tip of cerci level with ventral margin (Fig. 415) or slightly projecting beyond this. Aedeagus with medium-sized ventral plates. Styli large and forming an angle of 140-150° with longitudinal axis of aedeagus (Fig. 421). Vesica sclerotised and shining.

♀. Narrowest part of frons 0.33-0.36× head-width. Fore femoral organ present although sometimes only indicated. Mid femoral organ large (Fig. 424). Scutellum with 3 pairs of lateral bristles. Terminalia varying from black to red, most often reddish dorsally and blackish ventrally. T6 desclerotised dorsally, each half gradually narrowing medially, and with long bristly hairs posterior to the marginal bristles. Lateral marginal bristles of T6 stronger than dorsal (= median) marginals. T8 present as two sclerites, often weakly sclerotised but always distinctly setose. ST7 with several bristles in postero-lateral position (Fig. 423).

Length ♂♀. 8.0-18.0 mm.

176

Plate 1

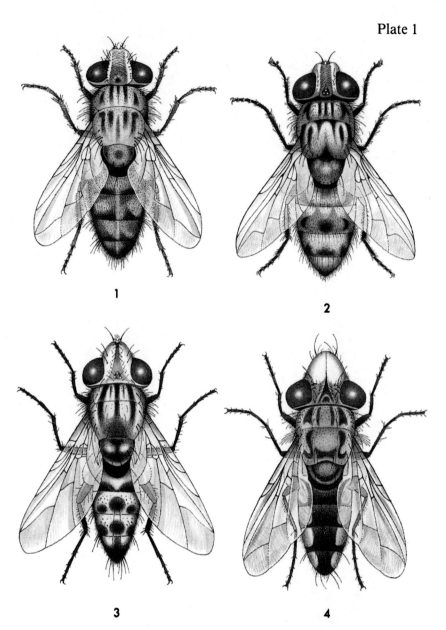

1

2

3

4

1: *Miltogramma oestraceum* (Fall.) ♂, × 7; 2: *Miltogramma punctatum* Meig. ♂, × 7; 3: *Taxigramma elegantula* (Zett.) ♂, × 9; 4: *Metopia argyrocephala* (Meig.) ♂, × 9. – Kuno Pape del.

Plate 2

1: *Macronychia striginervis* (Zett.) ♂, × 6; 2: *Blaesoxipha laticornis* (Meig.) ♂, × 7; 3: Same ♀, × 7; 4: *Sarcophaga carnaria* (L.) ♂, × 4. – Kuno Pape del.

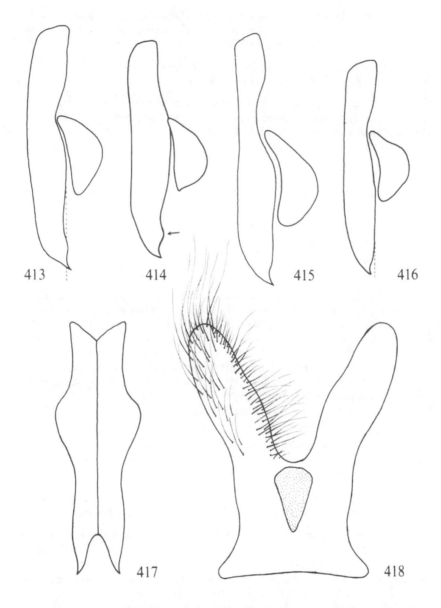

Figs 413-416. *Sarcophaga* spp., cerci + surstyli, lateral view. – 413: *S. variegata* (Scopoli); 414: *S. lasiostyla* Macquart; 415: *S. carnaria* (Linnaeus); 416: *S. subvicina* Rohdendorf.
Figs 417, 418. *Sarcophaga carnaria* (Linnaeus). – 417: cerci, posterior view; 418: ST5 ♂.

177

Distribution. Common in Denmark and Fennoscandia, but not reaching as far north as *S. variegata*. – From the British Isles and southern Europe east to the Altai Mountains; north to the Kola Peninsula (Verves 1981a).

Biology. Larvae predatory on earthworms, and adults attracted to faeces and decaying meat (Gregor & Povolný 1961; Grunin 1964).

Note. With the exception of *S. subvicina*, females of Fennoscandian and Danish species of *Sarcophaga* are at present inseparable, and I have been unable to confirm the differences mentioned by Kulikova (1982).

### 85. *Sarcophaga lasiostyla* Macquart, 1843
Figs 414, 420.

*Sarcophaga lasiostyla* Macquart, 1843, Hist. nat. Ins., Dipt., 2: 257.
*Sarcophaga cognata* Rondani, 1860, Atti Soc. ital. Sci. nat., 3: 385.
*Sarcophaga lehmanni* Müller, 1922, Arch. Naturgesch., 88A (2): 91.
*Sarcophaga carnaria meridionalis* Rohdendorf, 1937, Fauna USSR, Dipt., 19 (1): 284.

♂. Narrowest part of frons 0.17-0.22×head-width. Thorax, legs, and abdomen as in *S. carnaria*. Terminalia black. Protandrial segment non-pollinose or with faint traces of pollinosity, and with marginal bristles. Tip of cerci more or less level with ventral margin, or projecting somewhat beyond this. Subapical swelling of the anterior margin well-developed (Fig. 414). Aedeagus with well-developed ventral plates. Styli of medium size, not projecting beyond margin of ventral plates. Styli and juxta forming an angle of 95-110° with longitudinal axis of aedeagus (Fig. 420). Juxta somewhat more sclerotised than in *S. variegata*.
♀. Indistinguishable from the female *S. carnaria*.
Length ♂♀. 7.0-15.0 mm.

Distribution. Common in Denmark and southern Sweden and Norway. Not recorded from Finland. – Widely distributed from western Europe and North Africa to Central Asia (Verves 1980). Not in the British Isles.

Biology. Larvae predatory on earthworms, and as in *S. variegata* adults are attracted to meat and faeces (Viktorov-Nabokov & Verves 1979; Gregor & Povolný 1961).

Note. The marked similarity between *S. variegata* and *S. lasiostyla* has led many authors to consider the latter as a variety of the former (see Lehrer (1973b) for references). The two species are broadly sympatric, they share the same microhabitat, and they even seem to occupy the same niche, but the diagnostic characters separating them are reasonably constant throughout their range.
There seems to be a gradual transition in the relative frequency of *S. variegata* and *S. lasiostyla* from almost pure populations of *S. lasiostyla* in the warmer Mediterranean areas to pure populations of *S. variegata* in the north.

86. *Sarcophaga variegata* (Scopoli, 1763)
   Figs 413, 419.

*Musca variegata* Scopoli, 1763, Entom. carniolica: 326.
*Sarcophaga carnaria;* auctt., *nec* Linnaeus, 1758.

♂. Narrowest part of frons 0.20-0.24×head-width. Thorax, legs, and abdomen as in *S. carnaria*. Terminalia black. Protandrial segment non-pollinose, or with faint traces of pollinosity, and with marginal bristles. Tip of cerci projecting beyond ventral margin (Fig. 413). Aedeagus with well-developed ventral plates. Styli large, projecting be-

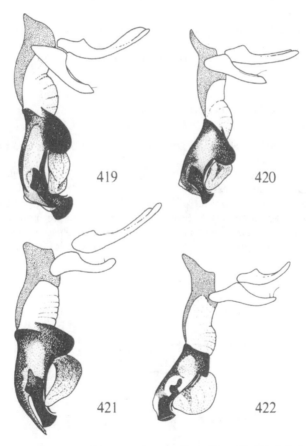

Figs 419-422. *Sarcophaga* spp., aedeagus. – 419: *S. variegata* (Scopoli); 420: *S. lasiostyla* Macquart; 421: *S. carnaria* (Linnaeus); 422: *S. subvicina* Rohdendorf.

179

yond margin of ventral plates. Styli and juxta forming an angle of 120-130° with longitudinal axis of aedeagus (Fig. 419).

♀. Indistinguishable from the female *S. carnaria*.

Length ♂♀. 6.0-18.0 mm.

Distribution: Very common in Denmark and Fennoscandia. Reaching as far north as TR in Norway. – Widely distributed in the Palaearctic region, from the British Isles to Kamchatka (Verves 1980).

Biology: Although often listed as breeding in a wide variety of decaying animal matter, causing wound and dermal myiasis, and as parasitising/preying upon snails and insects (e.g. James 1947; Séguy 1941; Emden 1954), the species actually seems to be an obligate predator of earthworms (Eberhardt & Steiner 1952; Kirchberg 1954, 1961). However, adults are often attracted to putrefying substrates (Kirchberg 1961; Gregor & Povolný 1961; Mihályi 1965; Hanski & Kuusela 1980), and larvae have been reared successfully from bird liver and horsemeat (Kirchberg 1954; Draber-Mońko 1973).

Bred from *Lumbricus terrestris* (Linnaeus) in Denmark.


### 87. *Sarcophaga subvicina* Rohdendorf, 1937
Figs 416, 422.

*Sarcophaga vicina* Villeneuve, 1899, Bull. Soc. ent. Fr., 1899: 27. Preocc. by Macquart, 1835.

*Sarcophaga subvicina* Rohdendorf, 1937, Fauna USSR, Dipt., 19 (1): 285. New name for *vicina* Villeneuve, 1899.

♂. Narrowest part of frons 0.18-0.23 × head-width. Thorax, legs, and abdomen as in *S. carnaria*. Terminalia black. Protandrial segment with grey pollinosity, varying from faint traces to a thin but distinct covering. Tip of cerci placed behind ventral margin (Fig. 416). Aedeagus with small ventral plates, large vesica, and short styli (Fig. 422).

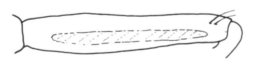

**424**

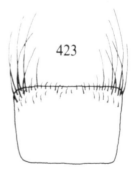

Figs 423, 424. *Sarcophaga* sp. ♀. –
423: ST7; 424: mid femoral organ.

♀. Very similar to other species of *Sarcophaga,* but abdominal T8 is completely absent (dissection necessary).

Length ♂♀. 8.0-15.0 mm.

Distribution. Common in Denmark and Fennoscandia, but not reaching as far north as *S. carnaria.* – Widely distributed in the Palaearctic region, from the British Isles east to the Urals and the Caucasus.

Biology. No records from natural conditions, but larvae have been bred in the laboratory on a dead slug (*Limax* sp.) (K. Rognes *in litt.* 1985), and on meat and liver (Pollock 1972; Baudet 1982; Blackith & Blackith 1984).

Note. Emden (1954) mentioned a difference in the hairing of the protandrial segment of *S. variegata* and *S. subvicina,* the latter possessing denser, longer, and more wavy hairs, but I have been unable to confirm this. It should be noted that significant infraspecific variation in hairiness can result from variations in food supply in the larval stage: small specimens are often less hairy.

| | | Germany | G. Britain | SJ | EJ | WJ | NWJ | NEJ | F | LFM | SZ | NWZ | NEZ | B | Sk. | Bl. |
|---|---|---|---|---|---|---|---|---|---|---|---|---|---|---|---|---|
| *Miltogramma brevipilum* Vill. | 1 | | | | | | | | | | | | | | | |
| *M. germari* Meig. | 2 | ● | ● | ● | ● | ● | ● | ● | ● | | | | ● | | ● | ● |
| *M. ibericum* Vill. | 3 | | | | | | | | | | | | | | | |
| *M. oestraceum* (Fall.) | 4 | ● | | ● | | | ● | | | ● | ● | ● | ● | ● | ● | |
| *M. punctatum* Meig. | 5 | ● | ● | ● | ● | ● | ● | ● | ● | | | | ● | ● | ● | ● |
| *M. testaceifrons* (v. Roser) | 6 | | | | | | | | | | | | | | | |
| *M. villeneuvei* Verves | 7 | | | | | | | | | | | | | | | |
| *Pterella grisea* (Meig.) | 8 | ● | ● | | | | | ● | | | | | ● | | | |
| *Senotainia conica* (Fall.) | 9 | ● | ● | ● | | ● | ● | | | ● | ● | | ● | ● | ● | ● |
| *S. puncticornis* (Zett.) | 10 | | | | | | | | | | | | | ● | ● | |
| *S. albifrons* (Rond.) | 11 | ● | | | | | | | | | | | | | | ● |
| *S. tricuspis* (Meig.) | 12 | | | | | | | | | | | | | | | ● |
| *Amobia signata* (Meig.) | 13 | ● | ● | | ● | | | | | ● | ● | ● | ● | | | |
| *A. oculata* (Zett.) | 14 | | | | | | | | | | | | | | | |
| *Phylloteles pictipennis* Lw. | 15 | | | | | | | | | | | | | | | |
| *Oebalia minuta* (Fall.) | 16 | ● | ● | | ● | ● | | ● | ● | | | ● | ● | | ● | |
| *O. cylindrica* (Fall.) | 17 | ● | ● | | ● | ● | | ● | ● | | | ● | ● | | | |
| *O. sachtlebeni* Rohd. | 18 | | | | ● | | | | ● | | | | ● | | | |
| *Hilarella hilarella* (Zett.) | 19 | ● | | | ● | | | | | | | | | | ● | ● |
| *H. stictica* (Meig.) | 20 | ● | | ● | ● | ● | | ● | ● | | | ● | ● | | ● | |
| *Taxigramma heteroneurum* (Meig.) | 21 | ● | | | | | | | | | | | ● | | | |
| *T. elegantulum* (Zett.) | 22 | ● | | | | ● | ● | ● | | | | | | | ● | ● |
| *Phrosinella sannio* (Zett.) | 23 | | | | | | | | | | | | | | | |
| *P. nasuta* (Meig.) | 24 | | | | | | | | | | | | | | | |
| *Metopia campestris* (Fall.) | 25 | ● | ● | ● | ● | ● | ● | | ● | ● | ● | ● | ● | ● | ● | ● |
| *M. grandii* Vent. | 26 | ● | | | | | ● | | | | | | | | | |
| *M. argyrocephala* (Meig.) | 27 | ● | ● | ● | ● | ● | ● | ● | ● | ● | | | ● | ● | ● | ● |
| *M. staegerii* Rond. | 28 | ● | | ● | ● | ● | ● | ● | ● | | ● | ● | ● | | | |
| *M. tshernovae* Rohd. | 29 | | | | ● | | | ● | ● | ● | | ● | ● | | | |
| *M. roserii* Rond. | 30 | | | | | | | | | | | | | | | |
| *Macronychia agrestis* (Fall.) | 31 | ● | | | ● | | | | | | | | | | ● | ● |
| *M. griseola* (Fall.) | 32 | ● | ● | | | ● | | | | | | | | ● | | ● |
| *M. polyodon* (Meig.) | 33 | ● | ● | | | | | | | | | | | | | ● |
| *M. striginervis* (Zett.) | 34 | ● | ● | | | | | | | | | | ● | | | ● |
| *Agria punctata* R.-D. | 35 | ● | ● | | ● | ● | | | | | | | ● | ● | | ● |
| *A. mamillata* (Pand.) | 36 | ● | ● | | | | | | | | | | ● | | | ● |
| *Sarcophila latifrons* (Fall.) | 37 | ● | ● | ● | ● | ● | ● | ● | ● | ● | | | ● | ● | ● | ● |
| *Angiometopa falleni* Pape | 38 | ● | ● | | ● | | | | | | | | ● | | ● | ● |

| | Hall. | Sm. | Öl. | Gtl. | G. Sand. | Ög. | Vg. | Boh. | Dlsl. | Nrk. | Sdm. | Upl. | Vstm. | Vrm. | Dlr. | Gstr. | Hls. | Med. | Hrj. | Jmt. | Äng. | Vb. | Nb. | Ås. Lpm. | Ly. Lpm. | P. Lpm. | Lu. Lpm. | T. Lpm. |
|---|---|---|---|---|---|---|---|---|---|---|---|---|---|---|---|---|---|---|---|---|---|---|---|---|---|---|---|---|
| 1 | | | ● | ● | | | | | | | | | | | | | | | | | | | | | | | | |
| 2 | ● | ● | ● | ● | | | ● | | | | | | | | | | | | | | | | | | | | | |
| 3 | | | | | | | | | | | | | | | | | | | | | | | | | | | | |
| 4 | | ● | ● | | | ● | | | | ● | | | | | | | | | | | | | | | | | | |
| 5 | ● | ● | | ● | | | | | | | | ● | | | | | | | | | | | | | | | | |
| 6 | | | | | | | | | | | | | | | | | | | | | | | | | | | | |
| 7 | | | | | | | | | | | | | | | | | | | | | | | | | | | | |
| 8 | | ● | | | | | | | | | | | | | | | | | | | | | | | | | | |
| 9 | ● | ● | | ● | | | | | | | ● | ● | | ● | | | | | | | | ● | | | | | | |
| 10 | | ● | | ● | | | | | | | ● | | | ● | | | | | | ● | | ● | | | ● | | | ● |
| 11 | | | | | | | | | | | | | | | | | | | | | | | | | | | | |
| 12 | | | | | | | | | | | | | | | | | | | | | | | | | | | | |
| 13 | | | | | | | | | | | | ● | | | | | | | | | | | | | | | | |
| 14 | | | ● | | | | | | | | | ● | | | | | | | | | | | | | ● | | | |
| 15 | | | | | | | | | | | | | | | | | | | | | | | | | | | | |
| 16 | | ● | | ● | | | | | | ● | ● | ● | | | | ● | | | | ● | ● | | | | | | | |
| 17 | | ● | | ● | | | | | | | | | ● | | | | | | ● | ● | | | | | | | ● | |
| 18 | | ● | ● | ● | | | | | | | | ● | | | | | | | | | | | | | | | | |
| 19 | | ● | ● | ● | | | | | | | | | | | | | | | | | | ● | | | | | ● | |
| 20 | | ● | ● | ● | | | | | | | | | | | | | | | | | | | | | | | | |
| 21 | | | ● | | | | | | | | | | | | | | | | | | | | | | | | | |
| 22 | | | | ● | | | | | | | | | | | | | | | | | | ● | | | | | | |
| 23 | | ● | | | | | | | | | | | | | | | | | | | | | | | | | ● | |
| 24 | | | | | | | | | | | | | | ● | | | | | | | | | | | | | | |
| 25 | ● | | ● | | | | | | | | ● | ● | | ● | ● | ● | | | ● | ● | | | | | | | | ● |
| 26 | | | | | | | | | | | | ● | | | | | | | | | | | | | | | | |
| 27 | ● | ● | ● | | | | | | | | ● | | ● | | | | | | | ● | | ● | | | | | | ● |
| 28 | | ● | | ● | | | | | | | | | | | | | | | | | | | | | | | | |
| 29 | | | | | | | | | | | | | | | | | | | | | | | | | | | | |
| 30 | | | | | | | | | | | | | | | | | | | | | | | | | | | | |
| 31 | | ● | ● | ● | | | ● | | | | | | | | | | | | | | | | | | | | | |
| 32 | | ● | | ● | | | | | | | | | | | ● | | | | | | | | | | | | | |
| 33 | | ● | ● | ● | | | | | | ● | | | | | | | | | | | | | | | | | | |
| 34 | | ● | | ● | | | | | | | | | | | | | | | | | | | | ● | | ● | | |
| 35 | | ● | ● | | | | | | | | | | | ● | | | | | | | | | | | | | | |
| 36 | | | ● | | | | | | | ● | | | | ● | | | | | | | | | | | | | | |
| 37 | | ● | ● | ● | | | | | | | | | | | | | | | | | | | | | | | | |
| 38 | ● | | ● | ● | | | | | | | | | | | | | | | | | | | | | | | | |

183

NORWAY

| | | Ø + AK | HE (s + n) | O (s + n) | B (ø + v) | VE | TE (y + i) | AA (y + i) | VA (y + i) | R (y + i) | HO (y + i) | SF (y + i) | MR (y + i) | ST (y + i) | NT (y + i) | Ns (y + i) |
|---|---|---|---|---|---|---|---|---|---|---|---|---|---|---|---|---|
| *Miltogramma brevipilum* Vill. | 1 | | ● | | | | | | | | | | | | | |
| *M. germari* Meig. | 2 | | | | | | | | | | | | | | | |
| *M. ibericum* Vill. | 3 | | | | | | | | | | | | | | | |
| *M. oestraceum* (Fall.) | 4 | | | | | | | | | | | | | | | |
| *M. punctatum* Meig. | 5 | | | | | | | | ● | | | | | | | |
| *M. testaceifrons* (v. Roser) | 6 | | | | | | | | | | | | | | | |
| *M. villeneuvei* Verves | 7 | | | | | | | | | | | | | | | |
| *Pterella grisea* (Meig.) | 8 | | | | | | | | | | | | | | | |
| *Senotainia conica* (Fall.) | 9 | ● | ● | ● | ● | ● | ● | ● | | | ● | | | | | |
| *S. puncticornis* (Zett.) | 10 | ● | ● | ● | | | | | | | ● | | ● | | | |
| *S. albifrons* (Rond.) | 11 | | | | | | | | | | | | | | | |
| *S. tricuspis* (Meig.) | 12 | | | | | | | | | | | | | | | |
| *Amobia signata* (Meig.) | 13 | | | | | | | | | | | | | | | |
| *A. oculata* (Zett.) | 14 | ● | | | | | ● | ● | | | | | | | | |
| *Phylloteles pictipennis* Lw. | 15 | | | | | | | | | | | | | | | |
| *Oebalia minuta* (Fall.) | 16 | ● | | | | | ● | ● | | | | | | | | |
| *O. cylindrica* (Fall.) | 17 | ● | | | | | | | | ● | | | | | | |
| *O. sachtlebeni* Rohd. | 18 | ● | | | | | | | | | | | | | | |
| *Hilarella hilarella* (Zett.) | 19 | | | | | | | | | | | | | | | |
| *H. stictica* (Meig.) | 20 | | | | | | | | | | | | | | | |
| *Taxigramma heteroneurum* (Meig.) | 21 | | | | | | | | | | | | | | | |
| *T. elegantulum* (Zett.) | 22 | | | | | | | | | | | | | | | |
| *Phrosinella sannio* (Zett.) | 23 | | ● | | | | ● | ● | | | | | | | ● | |
| *P. nasuta* (Meig.) | 24 | | | | | | | | | | | | | | | |
| *Metopia campestris* (Fall.) | 25 | ● | ● | ● | | | ● | | | ● | ● | ● | ● | ● | | |
| *M. grandii* Vent. | 26 | | | | | | | | | | | | | | | |
| *M. argyrocephala* (Meig.) | 27 | ● | | ● | ● | | ● | ● | | ● | ● | ● | ● | | | ● |
| *M. staegerii* Rond. | 28 | | ● | | | | | | | | | | | | | |
| *M. tshernovae* Rohd. | 29 | ● | ● | ● | | | | | | | | | | | | |
| *M. roserii* Rond. | 30 | | | | | | | | | | | | | | | |
| *Macronychia agrestis* (Fall.) | 31 | ● | ● | | | | | | | | | | | | | |
| *M. griseola* (Fall.) | 32 | | | | | | | | | | | | | | | |
| *M. polyodon* (Meig.) | 33 | | | | | | | | ● | | | | | | | |
| *M. striginervis* (Zett.) | 34 | | ● | | | | ● | | ● | | | | | | | |
| *Agria punctata* R.-D. | 35 | | | | | | | | | | | | | | | |
| *A. mamillata* (Pand.) | 36 | ● | | | | | ● | | | | | | | | | |
| *Sarcophila latifrons* (Fall.) | 37 | | | | | | | | | | | | | | | |
| *Angiometopa falleni* Pape | 38 | | | | | | | | | | | | | | | |

184

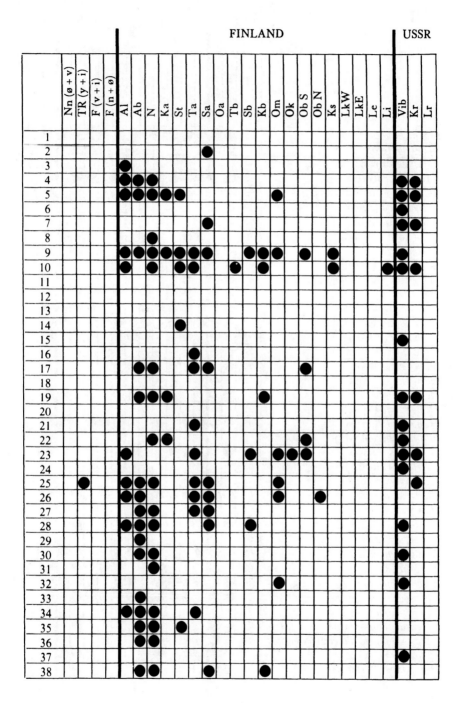

| | | Germany | G. Britain | SJ | EJ | WJ | NWJ | NEJ | F | LFM | SZ | NWZ | NEZ | B | Sk. | Bl. |
|---|---|---|---|---|---|---|---|---|---|---|---|---|---|---|---|---|
| *Brachicoma devia* (Fall.) | 39 | ● | ● | ● | ● | ● | ● | ● | ● | ● | ● | ● | ● | ● | ● | |
| *Nyctia halterata* (Pz.) | 40 | ● | ● | | | | | | ● | | | | | | | |
| *Paramacronychia flavip.* (Girsch.) | 41 | | | | | | | | | | | | | | ● | |
| *Ravinia pernix* (Harr.) | 42 | ● | ● | ● | ● | ● | ● | ● | ● | ● | | | ● | ● | ● | |
| *Blaesoxipha agrestis* (R.-D.) | 43 | ● | | | ● | | | | | | | | | | ● | |
| *B. pygmaea* (Zett.) | 44 | ● | | | | | | | | | | | | | ● | |
| *B. plumicornis* (Zett.) | 45 | ● | ● | | | | | | | | | | | | ● | |
| *B. laticornis* (Meig.) | 46 | ● | | ● | ● | | ● | ● | | ● | | ● | ● | ● | | |
| *B. rossica* Vill. | 47 | ● | ● | | ● | | ● | | ● | | | | ● | | ● | |
| *B. erythrura* (Meig.) | 48 | ● | ● | | | | | | | | | | | | | |
| *Bellieriomima subulata* (Pand.) | 49 | ● | ● | ● | | | | | | | | | | | ● | |
| *Discachaeta pumila* (Meig.) | 50 | ● | ● | | | | | ● | ● | ● | ● | ● | | | ● | |
| *Helicophagella melanura* (Meig.) | 51 | ● | ● | ● | ● | ● | ● | ● | | | | | ● | ● | ● | |
| *H. crassimargo* (Pand.) | 52 | ● | ● | | | | ● | ● | ● | ● | ● | ● | ● | ● | ● | |
| *H. agnata* (Rond.) | 53 | ● | ● | | ● | | | ● | ● | | | ● | | ● | | |
| *H. rosellei* (Böttcher) | 54 | ● | ● | | | | | | | | | | | | | |
| *H. noverca* (Rond.) | 55 | ● | | | | | | | | | | | | | | |
| *H. hirticrus* (Pand.) | 56 | | ● | | | | | | | | | | | | | |
| *Heteronychia depressifrons* (Zett.) | 57 | ● | ● | | | | ● | | ● | ● | | ● | ● | ● | ● | |
| *H. bezziana* (Böttcher) | 58 | | | | | | | | | | | | | | | |
| *H. haemorrhoa* (Meig.) | 59 | ● | ● | ● | ● | | ● | | ● | ● | ● | ● | ● | ● | ● | |
| *H. boettcheriana* (Rohd.) | 60 | | | | | | | | | | | | | ● | ● | |
| *H. vagans* (Meig.) | 61 | ● | ● | | ● | ● | | | ● | | | ● | ● | ● | ● | |
| *H. vicina* (Macq.) | 62 | | | | | | | | | | | | | | | |
| *H. proxima* (Rond.) | 63 | | | | | | | | | | | | | | | |
| *Pierretia sexpunctata* (F.) | 64 | ● | ● | ● | ● | ● | ● | ● | | ● | | ● | | ● | ● | |
| *P. nemoralis* (Kramer) | 65 | | | | | | | | | | | | | | | |
| *P. villeneuvei* (Böttcher) | 66 | ● | ● | | | | | | | | | | | ● | | ● |
| *P. socrus* (Rond.) | 67 | | | | | | | | | | | | | | | |
| *P. nigriventris* (Meig.) | 68 | ● | ● | ● | | | | ● | ● | | | ● | ● | | | |
| *P. soror* (Rond.) | 69 | ● | ● | | | | ● | | | | ● | | ● | ● | | |
| *Sarcotachinella sinuata* (Meig.) | 70 | ● | ● | ● | ● | ● | ● | ● | | ● | ● | | ● | ● | ● | |
| *Thyrsocnema incisilobata* (Pand.) | 71 | ● | ● | ● | ● | ● | ● | ● | ● | ● | ● | ● | ● | ● | ● | |
| *T. kentejana* Rohd. | 72 | | | | | | | | | | | | | | | |
| *Bercaea cruentata* (Meig.) | 73 | ● | ● | | | | | | | | | | ● | | | |
| *Parasarcophaga albiceps* (Meig.) | 74 | ● | ● | | | | | | | | | | | | ● | |
| *P. aratrix* (Pand.) | 75 | ● | ● | ● | ● | | ● | | | | | | ● | ● | ● | |
| *P. uliginosa* (Kramer) | 76 | ● | ● | ● | | | | | | | | | ● | | | |

186

SWEDEN

| | Hall. | Sm. | Öl. | Gtl. | G. Sand. | Ög. | Vg. | Boh. | Dlsl. | Nrk. | Sdm. | Upl. | Vstm. | Vrm. | Dlr. | Gstr. | Hls. | Med. | Hrj. | Jmt. | Ång. | Vb. | Nb. | Ås. Lpm. | Ly. Lpm. | P. Lpm. | Lu. Lpm. | T. Lpm. |
|---|---|---|---|---|---|---|---|---|---|---|---|---|---|---|---|---|---|---|---|---|---|---|---|---|---|---|---|---|
| 39 | ● | | ● | | | | | ● | ● | | | ● | | ● | ● | | | | ● | ● | | | | | | | ● | ● |
| 40 | | | | | | | | | | | | | | | | | | | | | | | | | | | | |
| 41 | | | | | | | | | | | | | | | | | | | | | | | | | | | | |
| 42 | | | ● | ● | | | ● | | | | | ● | | ● | | | | | | | | | | | | | | |
| 43 | | | ● | | | | ● | | | | | | | | | | | | | | | | | | | | | |
| 44 | | | | | | | | | | | | | | | | | | | | | | | | | | | | |
| 45 | | | ● | ● | | | | | | | | | | | | | | | | | | | | | | | | |
| 46 | | | ● | | | | | | | | | | | | | | | | | | | | | | | | | |
| 47 | | | ● | ● | | | | | | | | | | | | | | | | | | | | | | | | |
| 48 | | ● | | ● | | | | | | | | | | | | | | | | | | | | | | | | |
| 49 | | | | | | | | | | | | | | | | | | | | ● | ● | | | | | | | |
| 50 | | | | | | | | | | | | ● | | | | | | | | | | | | | | | | |
| 51 | ● | ● | ● | ● | | | ● | ● | | | | ● | | ● | | | | | | ● | ● | | | | | | | |
| 52 | ● | ● | ● | ● | | | | | | | | ● | | | | | | | | ● | ● | ● | ● | ● | | | | |
| 53 | | ● | | | | | | | | | | | | | | | | | | | | | | | | | | |
| 54 | | | | | | | | | | | | | | | | | | | ● | ● | ● | ● | | | | | ● | ● |
| 55 | | | | | | | | | | | | | | | | | | | | | | | | | | | | |
| 56 | | | ● | | | | | | | | | | | | | | | | ● | | | | | | | | | |
| 57 | | | | | | | ● | | | | ● | ● | ● | ● | | | | | | | | | | | | | | |
| 58 | | | | | | | | | | | | | | ● | | | | | | | | | | | | | | |
| 59 | ● | | ● | ● | | | | | ● | | | ● | | | | | | | | | | | | | | | | |
| 60 | | | ● | | | | ● | | | | | ● | | | | | | | | | | | | | | | | |
| 61 | | ● | | ● | | | | | | | | | | | | | | | ● | ● | ● | | | | ● | ● | ● | ● |
| 62 | | | | | | | | | | | | | | | | | | | | ● | ● | | | | | | ● | ● |
| 63 | | | ● | | | | | | | | | | | | | | | | | | | | | | | | | |
| 64 | | ● | ● | | | | | ● | | | ● | ● | | ● | ● | | | | | ● | | | | | | | ● | |
| 65 | | | | | | | | | | | | ● | | | | | | | | ● | | | | | | | | |
| 66 | | ● | ● | | | | | | | | | | | | | | | | ● | ● | | | | | | | | |
| 67 | | | | | | | | | | | | | | | | | | | | | | | | | | | | |
| 68 | | | | | | | | | | | | | | | | | | | | | | | | | | | | |
| 69 | | | | ● | | | | | | | | ● | | | | | | | | | | | | | | | | |
| 70 | ● | ● | | ● | | | | | | | | | | ● | | | | | | | | | | | | | ● | |
| 71 | ● | ● | ● | ● | | | | ● | ● | | | | | ● | | | | | | | | | | | | | | |
| 72 | | | | | | | | | | | | | | | | | | | ● | ● | | | | | ● | ● | | ● |
| 73 | | | | | | | | | | | | | | | | | | | | | | | | | | | | |
| 74 | | ● | | | | | | | | | | ● | | | | | | | | | | | | | | | | |
| 75 | | ● | ● | | | | ● | | | | | | | | | | | | | | | ● | | | | | | |
| 76 | | | | | | | | | | | | | | | | | | | | | | | | | | | | |

| | | Ø + AK | HE (s + n) | O (s + n) | B (ø + v) | VE | TE (y + i) | AA (y + i) | VA (y + i) | R (y + i) | HO (y + i) | SF (y + i) | MR (y + i) | ST (y + i) | NT (y + i) | Ns (y + i) |
|---|---|---|---|---|---|---|---|---|---|---|---|---|---|---|---|---|
| *Brachicoma devia* (Fall.) | 39 | ● | ● | ● | ● | ● | ● | | ● | ● | ● | | | ● | | |
| *Nyctia halterata* (Pz.) | 40 | | | | | | | | | | | | | | | |
| *Paramacronychia flavip.* (Girsch.) | 41 | | | | | | | | | | | | | | | |
| *Ravinia pernix* (Harr.) | 42 | ● | ● | ● | ● | ● | | | ● | | | | | | | |
| *Blaesoxipha agrestis* (R.-D.) | 43 | | | ● | ● | | | | | | | | | | | |
| *B. pygmaea* (Zett.) | 44 | | | | | | | | | | | | | | | |
| *B. plumicornis* (Zett.) | 45 | ● | | ● | | ● | | ● | | | ● | | | | | |
| *B. laticornis* (Meig.) | 46 | | | | | | | | | | | | | | | |
| *B. rossica* Vill. | 47 | | | | | | | | | | ● | | | | | |
| *B. erythrura* (Meig.) | 48 | | | | | | | | | | | | | | | |
| *Bellieriomima subulata* (Pand.) | 49 | ● | | | | | | | | | | | | | | |
| *Discachaeta pumila* (Meig.) | 50 | ● | | ● | | | | | | | | | | | | |
| *Helicophagella melanura* (Meig.) | 51 | ● | ● | | ● | | | ● | ● | ● | | ● | | | | |
| *H. crassimargo* (Pand.) | 52 | ● | ● | ● | | | | | ● | | | | | | | |
| *H. agnata* (Rond.) | 53 | | | | | | | ● | ● | ● | | | | | | |
| *H. rosellei* (Böttcher) | 54 | ● | | ● | | | ● | | ● | | ● | | ● | | | ● |
| *H. noverca* (Rond.) | 55 | | | | | | | | | | | | | | | ● |
| *H. hirticrus* (Pand.) | 56 | | | | | | | | | | | | | | | |
| *Heteronychia depressifrons* (Zett.) | 57 | ● | | | | | ● | | | | | | | | | |
| *H. bezziana* (Böttcher) | 58 | ● | | | | | | | | | | | | | | |
| *H. haemorrhoa* (Meig.) | 59 | ● | | | | | ● | ● | | | | | | | | |
| *H. boettcheriana* (Rohd.) | 60 | ● | | | | | ● | | | | | | | | | |
| *H. vagans* (Meig.) | 61 | ● | ● | ● | ● | ● | ● | | ● | | ● | | ● | ● | | ● |
| *H. vicina* (Macq.) | 62 | ● | ● | ● | | | | | ● | ● | | ● | ● | ● | | ● |
| *H. proxima* (Rond.) | 63 | | | | | | | | | | | | | | | |
| *Pierretia sexpunctata* (F.) | 64 | | ● | | | | ● | ● | | ● | | ● | | | | |
| *P. nemoralis* (Kramer) | 65 | ● | | | | | | | | | | | | | | |
| *P. villeneuvei* (Böttcher) | 66 | | | | | | | | | | | | | | | |
| *P. socrus* (Rond.) | 67 | | | | | | | | | | | | | | | |
| *P. nigriventris* (Meig.) | 68 | | | | | | | | | | | | | | | |
| *P. soror* (Rond.) | 69 | ● | | ● | | | | | | | | | | | ● | |
| *Sarcotachinella sinuata* (Meig.) | 70 | ● | | | | | | | ● | | | | | | | |
| *Thyrsocnema incisilobata* (Pand.) | 71 | ● | | ● | ● | ● | ● | | ● | ● | ● | | | | | |
| *T. kentejana* Rohd. | 72 | | | | | | ● | | | | | | | | | ● |
| *Bercaea cruentata* (Meig.) | 73 | | | | | | | | | | | | | | | |
| *Parasarcophaga albiceps* (Meig.) | 74 | ● | | | | | | | | | | | | | | |
| *P. aratrix* (Pand.) | 75 | ● | | ● | | | ● | ● | ● | | ● | | | | | |
| *P. uliginosa* (Kramer) | 76 | | | | | | | | | | | | | | | |

| | Nn(ø+v) | TR(y+i) | F(v+i) | F(n+ø) | Al | Ab | N | Ka | St | Ta | Sa | Oa | Tb | Sb | Kb | Om | Ok | ObS | ObN | Ks | LkW | LkE | Lc | Li | Vib | Kr | Lr |
|---|---|---|---|---|---|---|---|---|---|---|---|---|---|---|---|---|---|---|---|---|---|---|---|---|---|---|---|
| 39 | | ● | ● | ● | ● | ● | ● | | | ● | ● | | ● | ● | | | | | | | | | | | ● | ● | |
| 40 | | | | | | | | | | | | | | | | | | | | | | | | | | | |
| 41 | | | | | | | | | | | | | | | | | | | | | | | | | | | |
| 42 | | | | | ● | ● | ● | | ● | ● | | | | ● | ● | ● | | | ● | ● | ● | | | | ● | ● | |
| 43 | | | | | | | | | | | | | | | | | | | | | | | | | | | |
| 44 | | | | | | | | | | | | | | | | | | | | | | | | | | | |
| 45 | | | | | ● | ● | ● | | | | | | | | | | | | | | | | | | | | ● |
| 46 | | | | | | | | | | | | | | | | | | | | | | | | | | | ● |
| 47 | | | | | | | | | | | | | | | | | | | | | | | | | | | |
| 48 | | | | | ● | ● | ● | | | ● | ● | | | | | | | | ● | | | | | | | | ● |
| 49 | | | | | ● | ● | | | | | | | | | | | | | | | | | | | | | |
| 50 | | | | | | | ● | | | | ● | | | | | | | | | | | | | | | | |
| 51 | | | | | ● | ● | ● | | | ● | ● | ● | | ● | ● | ● | | | | | | | | | | ● | |
| 52 | ● | | | | ● | ● | ● | | | ● | ● | ● | ● | | | | ● | | | | | | | | ● | ● | |
| 53 | | | | | | | | | | | | | | | | | | | | | | | | | | | |
| 54 | ● | ● | ● | | | | | | | | | | | | | | | | | | | | | | | | |
| 55 | | | | | | | | | | | | | | | | | | | | | | | | | | | |
| 56 | | | | | | | | | | | | | | | | | | | | | | | | | | | |
| 57 | | | | | ● | ● | | | | | | | | | | | | | | | | | | | | | |
| 58 | | | | | | | | | | | | | | | | | | | | | | | | | | | |
| 59 | | | | | ● | ● | ● | | | | | | | | | | | | | | | | | | | | |
| 60 | | | | | | | | | | | | | | | | | | | | | | | | | | | |
| 61 | | ● | ● | | ● | ● | ● | | ● | | | | | ● | | | | | | | | | | | | ● | |
| 62 | | ● | | | ● | | | | | | | | | | | | | | | | | | | | | | |
| 63 | | | | | | | ● | | | | | | | | | | | | | | | | | | | ● | |
| 64 | | | | | | ● | ● | | | | ● | ● | | | | | | | | | | | ● | | ● | ● | |
| 65 | | | | | | | | | ● | | | | | | | | | | | | | ● | | | | | |
| 66 | | | | | | ● | ● | | | | | | | | | | | | | | | | | | | | |
| 67 | | | | | ● | | | | | ● | | | | | | | | | | | | | | | | ● | |
| 68 | | | | | | | | | | | | | | | | | | | | | | | | | | | |
| 69 | | | | | | | | | | | | | | | | | | | | | | | | | | | |
| 70 | | | | | ● | ● | ● | | ● | | | | | | | ● | ● | | | | | | | | | | |
| 71 | | | | | ● | ● | ● | ● | | | | | | | | | | | | | | | | | ● | ● | |
| 72 | | | | | | | | | | | | | | | | | | | | | | | | ● | | | |
| 73 | | | | | | | | | | | | | | | | | | | | | | | | | | | |
| 74 | | | | | | ● | ● | | | ● | ● | | | ● | ● | | | | | | | | | | | ● | |
| 75 | | | | | ● | ● | ● | | | ● | ● | | | | | | | ● | ● | | | | | | | | |
| 76 | | | | | | | | | | | | | | | | | | | | | | | | | | | |

189

# DENMARK

| | | Germany | G. Britain | SJ | EJ | WJ | NWJ | NEJ | F | LFM | SZ | NWZ | NEZ | B | Sk. | Bl. |
|---|---|---|---|---|---|---|---|---|---|---|---|---|---|---|---|---|
| *Parasarcophaga caerulescens* (Zett.) | 77 | ● | ● | ● | ● | ● | ● | ● | ● | ● | | ● | ● | | ● | |
| *P. jacobsoni* Rohd. | 78 | | | | | | | ● | | | | | | | | |
| *P. emdeni* Rohd. | 79 | | | | ● | | | | | | | | ● | ● | | |
| *P. portschinskyi* Rohd. | 80 | | | | ● | | | | | | | ● | ● | ● | | |
| *P. pleskei* Rohd. | 81 | | | | | | | | | | | | | | | |
| *P. similis* (Meade) | 82 | ● | ● | | | | | | | | | | | | | |
| *P. argyrostoma* (R.-D.) | 83 | ● | ● | | | | | | | | | | ● | | | |
| *Sarcophaga carnaria* (L.) | 84 | ● | ● | ● | | | | | ● | ● | ● | ● | | ● | ● | |
| *S. lasiostyla* Macq. | 85 | ● | ● | ● | ● | ● | ● | ● | | | | ● | ● | ● | ● | |
| *S. variegata* (Scop.) | 86 | ● | ● | ● | ● | ● | ● | ● | ● | ● | ● | ● | ● | ● | ● | ● |
| *S. subvicina* Rohd. | 87 | ● | ● | ● | ● | ● | ● | ● | | | | ● | ● | ● | ● | |

# NORWAY

| | | Ø+AK | HE (s+n) | O (s+n) | B (ø+v) | VE | TE (y+i) | AA (y+i) | VA (y+i) | R (y+i) | HO (y+i) | SF (y+i) | MR (y+i) | ST (y+i) | NT (y+i) | Ns (y+i) |
|---|---|---|---|---|---|---|---|---|---|---|---|---|---|---|---|---|
| *Parasarcophaga caerulescens* (Zett.) | 77 | ● | ● | ● | ● | | ● | ● | | | | | | | | |
| *P. jacobsoni* Rohd. | 78 | | | | | | | | | | | | | | | |
| *P. emdeni* Rohd. | 79 | ● | | | | | ● | | | | | | | | | |
| *P. portschinskyi* Rohd. | 80 | | | | | | ● | | ● | | | | | | | |
| *P. pleskei* Rohd. | 81 | | ● | | | | | | | | | | | | | |
| *P. similis* (Meade) | 82 | ● | | | | | | ● | | | | | | | | |
| *P. argyrostoma* (R.-D.) | 83 | | | | | | | | | | | | | | | |
| *Sarcophaga carnaria* (L.) | 84 | ● | ● | | ● | | ● | ● | ● | ● | ● | | | | | |
| *S. lasiostyla* Macq. | 85 | ● | | ● | ● | | | | | | | | | | | |
| *S. variegata* (Scop.) | 86 | ● | ● | ● | ● | ● | ● | ● | ● | ● | ● | ● | | | | ● |
| *S. subvicina* Rohd. | 87 | ● | | ● | ● | | ● | | ● | ● | ● | | ● | | ● | ● |

SWEDEN

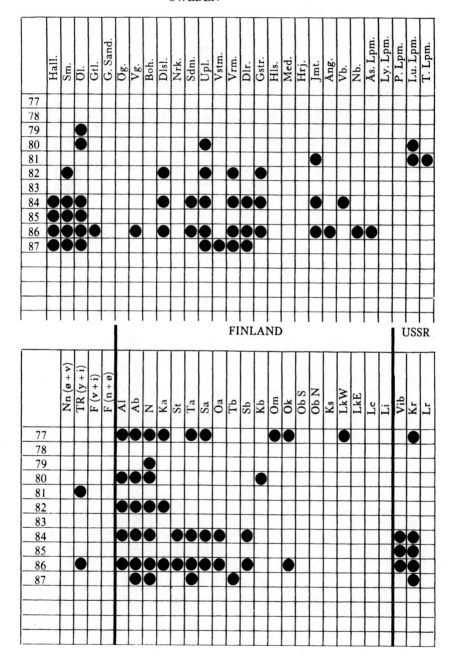

FINLAND          USSR

# Literature

Aldrich, J. M., 1916: *Sarcophaga* and allies in North America. – La Fayette, Indiana, 301 pp. + postscript 1 p. and index 3 pp., 16 pl.

Alford, D. V., 1975: Bumblebees. – Davis-Poynter, London, xii + 352 pp.

Allen, H. W., 1926: North American species of two-winged flies belonging to the tribe Miltogrammini. – Proc. U. S. natn. Mus. 68 (9): 1-106.

Aradi, P. & Mihályi, F., 1971: Seasonal investigations of flies visiting food markets in Budapest. – Acta zool. hung. 17 (1-2): 1-10.

Arnaud, P. H., Jr., 1954: *Hilarella hilarella* (Zetterstedt) (Diptera: Sarcophagidae) parasitic upon a rhaphidophorid (Orthoptera: Gryllacrididae). – Can. Ent. 86: 135-136.

Assis-Fonseca, E. C. M., 1953: An important character in the identification of the females of *Sarcophaga* species (Dipt., Calliphoridae). – J. Soc. Br. Ent. 4 (8): 167-168.

Audcent, H. L. F., 1951: Midnight flies. – Entomologist's mon. Mag. 87: 133.

Baranov, N. & Jezic, J., 1928: Fliegenmaden als Wundschmarotzer bei den Haustieren in Südserbien. – Z. ParasitKde 1: 416-422.

Barfoot, S. D., 1969: *Sarcophaga nigriventris* Meigen and *S. hirticrus* Pandellé (Dipt., Calliphoridae) both bred from *Helix aspersa* Müller (Mollusca, Helicidae). – Entomologist's mon. Mag. 105: 144.

Baudet, J. L., 1982: Contribution à la faunistique régionale du genre *Sarcophaga* (insectes diptères); critères de reconnaissance des femelles inventoriées. – Bull. Soc. Sci. nat. Ouest Fr. (N. S.) 4: 134-144.

Beaver, R. A., 1972: Ecological studies on Diptera breeding in dead snails. 1. Biology of the species found in *Cepaea nemoralis* (L.). – Entomologist 105: 41-52.

– 1977: Non-equilibrium "island" communities: Diptera breeding in dead snails. – J. Anim. Ecol. 46: 783-798.

– 1979: Biological studies of the fauna of pitcher plants (*Nepenthes*) in West Malaysia. – Annls Soc. ent. Fr. (N. S.) 15 (1): 3-17.

Blackith, R. & Blackith, R., 1984: Larval aggression in Irish flesh-flies (Diptera: Sarcophagidae). – Ir. Nat. J. 21 (6): 255-257.

Boiko, A. K., 1939: Larva of *Senotainia triguspis* (sic!) Meig. causing heavy losses of bees. – Dokl. Akad. Nauk SSSR 24 (3): 304-306.

– 1948: A new kind of myiasis in bumblebees. – Dokl. Akad. Nauk SSSR 61: 423-424. (In Russian).

Böttcher, G., 1912a: Zu Meigens und Pandellés *Sarcophaga*-Typen nebst Anmerkungen zu Kramers Tachiniden der Oberlausitz. (Dipt.). – Dt. ent. Z. 1912: 343-350.

– 1912b-1913: Die männlichen Begattungswerkzeuge bei dem Genus *Sarcophaga* Meig. und ihre Bedeutung für die Abgrenzung der Arten. – Dt. ent. Z. 1912 (5): 525-544, (6): 705-736; 1913 (1): 1-16, (2): 115-130, (3): 239-254, (4): 351-377.

Bowell, E. W., 1917: Larva of a dipterous fly feeding on *Helicella itala*. – Proc. malac. Soc. Lond. 12: 308.

Brauer, F. & Bergenstamm, J., 1889: Die Zweiflügler des Kaiserlichen Museums zu Wien. IV. Vorarbeiten zu einer Monographie der Muscaria Schizometopa (exclusive Anthomyidae). Pars I. – Denkschr. Akad. Wiss. Wien. Kl. math.-naturw. 56: 69-180.

Burgess, N. R., 1966: A case of human myiasis in London. – Trans. R. Soc. trop. Med. Hyg. 60 (4): 432-433.

Cameron, R. H. D. & Disney, R. H. L., 1975: Two further cases of parasitism by *Sarcophaga nigriventris* Meigen (Dipt., Sarcophagidae). - Entomologist's mon. Mag. 111: 45.

Catts, E. P., 1964: Field behavior of adult *Cephenemyia* (Diptera: Oestridae). - Can. Ent. 96: 579-585.

Chong, M., 1968: Notes and exhibitions: Releases of beneficial insects. - Proc. Hawaii. ent. Soc. 20 (1): 3.

Clausen, C. P., 1940: Entomophagous insects. - McGraw-Hill, x + 688 pp.

Day, C. D. & Assis-Fonseca, E. C. M., 1955: A key to the females of the British species of *Sarcophaga* (Dipt., Calliphoridae). - J. Soc. Br. Ent. 5 (4): 119-123.

Day, M. C. & Smith, K. G. V., 1980: Insect eggs on adult *Rhopalum clavipes* (L) (Hymenoptera: Sphecidae): a problem solved. - Entomologist's Gaz. 31 (3): 173-176.

De Geer, C., 1776: Mémoires pour servir à l'histoire des insectes. 6. Stockholm, viii + 523 pp.

Dodge, H. R., 1955: Sarcophagid flies parasitic on reptiles (Diptera, Sarcophagidae). - Proc. ent. Soc. Wash. 57: 183-187.

Downes, W. L., Jr., 1955: Notes on the morphology and classification of the Sarcophagidae and other calyptrates (Diptera). - Proc. Iowa Acad. Sci. 62: 514-538.

- 1965: Family Sarcophagidae. - Pp. 933-961 in Stone, A. et al. (Eds): A catalog of the Diptera of America north of Mexico. Agricultural Handbook No. 276, Washington, D. C.

Draber-Mońko, A., 1966: Bemerkungen über die paläarktischen Arten der Gattung *Pachyophthalmus* B. B. (Diptera, Sarcophagidae). - Polskie Pismo ent. 36: 395-405.

- 1973: Einige Bemerkungen über die Entwicklung von *Sarcophaga carnaria* (L.) (Diptera, Sarcophagidae). - Polskie Pismo ent. 43: 301-308, 5 pl.

- 1978: Scatophagidae, Muscinae, Gasterophilidae, Hippoboscidae, Calliphoridae, Sarcophagidae, Rhinophoridae, Oestridae, Hypodermatidae und Tachinidae (Diptera) der Pieninen. - Fragm. faun. 22 (2): 51-229. (In Polish with Russian and German summaries).

Eberhardt, A. I., 1955: Untersuchungen über das Schmarotzen von *Sarcophaga carnaria* an Regenwürmern und Vergleich der Biologie einiger *Sarcophaga* Arten. - Z. Morph. Ökol. Tiere 43: 616-647.

- & Steiner, G., 1952: Untersuchungen über das Schmarotzen von *Sarcophaga* spp. in Regenwürmern. - Z. Morph. Ökol. Tiere 41: 147-160.

Emden, F. I. van, 1950: Dipterous parasites of Coleoptera. - Entomologist's mon. Mag. 86: 182-206.

- 1954: Diptera Cyclorrhapha Calyptrata (1), section (a). Tachinidae and Calliphoridae. - Handbk Ident. Br. Insects 10, 4 (a): 1-133.

Enderlein, G., 1928: Klassifikation der Sarcophagiden. Sarcophagiden-Studien I. - Arch. klassif. phylogen. Ent. 1: 1-56.

Endo, A., 1980: The behaviour of a Miltogrammine fly *Metopia sauteri* (Townsend) (Diptera, Sarcophagidae) cleptoparasitizing on a spider wasp *Episyron arrogans* (Smith) (Hymenoptera, Pompilidae). - Kontyû 48 (4): 445-457.

Evans, H. E., 1970: Ecological-behavioral studies of the wasps of Jackson Hole, Wyoming. - Bull. Mus. comp. Zool. Harv. 140 (7): 451-511.

- & Eberhard, M. J. W., 1970: The wasps. - Univ. Michigan Press, Ann Arbor, 265 pp.

Fabricius, J. C., 1787: Mantissa insectorum sistens species nuper detectas adiectis synonymis, observationibus, descriptionibus, emendationibus. 2. - Hafniæ, 382 pp.

- 1794: Entomologia systematica emendata et aucta. Secundum classes, ordines, genera, species adjectis synonimis, locis, observationibus, descriptionibus. 4. Hafniæ, 472 pp.

- 1805: Systema antliatorum secundum ordines, genera, species adiectis synonymis, locis, observationibus, descriptionibus. - Brunsvigæ, 402 pp.

Fahlander, K., 1954: Hymenoptera från Gästrikland. - Ent. Tidskr. 75: 249-254.

Fallén, C. F., 1810: Försök att bestämma de i Sverige funne flugarter, som kunna föras till slägtet *Tachina.* – K. svenska VetenskAkad. Handl. [2] 31: 253-287.

– 1817: Beskrifning öfver de i Sverige funna fluge arter, som kunna föras till slägtet *Musca.* Första afdelningen. – K. svenska VetenskAkad. Handl. [3] 1816: 226-257.

– 1820: Monographia Muscidum Sveciae, Part 1: 1-12, Lundæ.

Ferton, C., 1901: Notes détachées sur l'instinct des Hyménoptères mellifères et ravisseurs avec la description de quelques espèces. – Annls Soc. ent. Fr. 70: 83-148.

Forsius, R., 1924: Über eine Massenzucht von *Hyponomeuta padi* L. – Notul. ent. 4: 44-46.

Forsyth, A. B. & Robertson, R. J., 1975: K reproductive strategy and larval behavior of the pitcher plant sarcophagid fly, *Blaesoxipha fletcheri.* – Can. J. Zool. 53: 174-179.

Girschner, E., 1881: Dipterologische Studien. – Ent. Nachricht. 7: 277-279.

Gregor, F. & Povolný, D., 1961: Resultate stationäre Untersuchungen von synanthropen Fliegen in der Umgebung einer Ortschaft in der Ostslowakei. – Zool. Listy 10: 17-44.

Griffiths, G. C. D., 1972: The phylogenetic classification of Diptera Cyclorrhapha with special reference to the structure of the male postabdomen. – Junk Publ., The Hague, 340 pp.

Grunin, K. Ya., 1964: On the biology and distribution of certain Sarcophaginae (Diptera, Sarcophagidae) in the USSR. – Ent. Rev., Wash. 43: 36-39. (Translated from Ént. Obozr. 43: 71-79).

Guilhon, J., 1945: Un nouveau cas d'apimyose. – Bull. Acad. vét. Fr. 18: 1-3.

Hackman, W., 1980: A check list of the Finnish Diptera II. Cyclorrhapha. – Notul. ent. 60: 117-162.

Hafez, M., 1940: A study of the morphology and life-history of *Sarcophaga falculata.* – Bull. Soc. Fouad I Ent. 24: 183-212.

Hanski, I. & Kuusela, S., 1980: The structure of carrion fly communities: differences in breeding seasons. – Ann. Zool. Fennici 17: 185-190.

Hardy, D. E., 1981: Diptera: Cyclorrhapha 4. – Insects Hawaii 14: 1-419.

Harris, M., 1776-1780: An Exposition of English insects, with curious observations and remarks, wherein each insect is particularly described; its parts and properties considered; the different sexes distinguished, and the natural history faithfully related. The whole illustrated with copper plates, drawn, engraved, and coloured, by the author. – London, viii + 9-166.

Hasselrot, T. B., 1960: Studies on Swedish bumblebees (genus *Bombus* Latr.), their domestication and biology. – Opusc. ent. Suppl. 17: 1-192, 10 pl.

Herting, B. & Simmonds, F. J., 1976: A catalogue of parasites and predators of terrestrial arthropods. Sect. A, host or prey/enemy. 7, Lepidoptera part 2 (Macrolepidoptera). – Commonwealth Agricultural Bureaux, 221 pp.

James, M. T., 1947: The flies that cause myiasis in man. – U.S. Dept. Agric. Misc. Publ. 631: 1-175.

Johnston, T. H. & Tiegs, C. W., 1921: New and little known Sarcophagidae flies from South Queensland. – Proc. R. Soc. Qd 33: 46-90.

Junnikkala, E., 1960: Life history and insect enemies of *Hyponomeuta malinellus* Zell. (Lep., Hyponomeutidae) in Finland. – Ann. Soc. zool.-bot. fennicae, Vanamo 21 (1): 1-44.

Kano, R., Field, G. & Shinonaga, S., 1967: Sarcophagidae (Insecta: Diptera). – Fauna japonica 12: 1-168, 41 pl.

Keilin, D., 1919: On the life-history and larval anatomy of *Melinda cognata* Meigen (Diptera Calliphorinae) parasitic in the snail *Helicella (Heliomanes) virgata* da Costa, with an account of the other Diptera living upon molluscs. – Parasitology 11: 430-455.

Kirchberg, E., 1954: Zur Larvennahrung einiger heimischer *Sarcophaga*-Arten, insbesondere zur Frage, ob *S. carnaria* L. als obligatorischer Regenwurmparasit anzusehen sei (Diptera, Tachinidae). – Z. Morph. Ökol. Tiere 43: 99-112.

Kleine, R., 1910: *Sarcophaga albiceps* Meig., Primärparasit bei *Saperda populnea* L. – Entomol. Blätt. 6 (9): 217-221.

194

Knipling, E. F., 1936: A comparative study of the first-instar larvae of the genus *Sarcophaga* (Calliphoridae, Diptera), with notes on biology. – J. Parasit. 22: 417-454.

Kramer, H., 1908: *Sarcophaga affinis* Fll. und Verwandte. – Ent. Wbl. 25: 200-201.
- 1917: Die Musciden der Oberlausitz. – Abh. naturforsch. Ges. Görlitz 27: 117-166.
- 1920: Zwei neue Deutsche Musciden. – Zool. Jb. Syst. 43: 329-332.

Kulikova, N. A., 1982: Utilization of ovipositor morphology for identification of female flies (Diptera, Sarcophagidae). – Zool. Zh. 61: 1518-1523. (In Russian with English summary).

Kurahashi, H., 1972: Studies on the calypterate muscoid flies from Japan IX. Subfamily Macronychiinae (Diptera, Sarcophagidae). – Kontyû 40 (3): 173-180.
- 1974: Note on the genus *Amobia* from the Indo-australian area with description of a new species (Diptera, Sarcophagidae). – Pacif. Insects 16 (1): 57-60.
- 1975: Studies on the calypterate muscoid flies from Japan XI. Subfamily Agriinae (Diptera, Sarcophagidae). – Kontyû 43 (2): 202-213.

Kuusela, S. & Hanski, I., 1982: The structure of carrion fly communities: the size and the type of carrion. – Holarct. Ecol. 5: 337-348.

Lehrer, A. Z., 1967: Espèces nouvelles du genre *Sarcophaga* Meigen (Fam. Sarcophagidae, Diptera). – Zool. Anz. 178 (3-4): 210-219.
- 1973a: Microsternites préabdominaux des Diptères cylindromyiides. – Zool. Anz. 190 (5-6): 405-408.
- 1973b: La taxonomie du genre *Sarcophaga* Meigen (Fam. Sarcophagidae, Diptera). – Annotnes zool. bot., Bratislava 89: 1-22.

Léonide, J., 1964: Contribution à l'étude biologique des diptères sarcophagidés, parasites d'acridiens: ponte de larve et infestation de l'hôte par le *Blaesoxipha berolinensis* Vill. – C. r. hebd. Séanc. Acad. Sci., Paris 258: 4352-4354.
- 1967: Contribution à l'étude biologique des diptères sarcophagidés, parasites d'acridiens. Cycle biologique de *Blaesoxipha rossica* Vill., injection de larves dans le corps de l'hôte par les femelles de sarcophagidés. – C. r. hebd. Séanc. Acad. Sci., Paris 265: 232-234.
- & Léonide, J.-C., 1971: Contribution à l'étude des diptères sarcophagidés acridiophages, V: Notes faunistique et biologique. – Bull. Soc. ent. Fr. 76: 111-122.
- 1975: Contribution à l'étude des diptères sarcophagidés acridiophages. X: Bio-taxonomie de *Blaesoxipha berolinensis* Villeneuve, 1912. – Bull. Soc. ent. Fr. 80: 6-19.
- 1979: Contribution à l'étude des diptères sarcophagidés acridiophages, XII. Biotaxonomie de *Blaesoxipha gladiatrix* (Pandellé 1896) Villeneuve 1911. – Bull. Soc. ent. Fr. 84: 247-265.
- 1982: Contribution à l'étude des diptères sarcophagidés acridiophages. XIII. – Biotaxonomie de *Blaesoxipha grylloctona* Loew 1861. Nouvelles réflection sur la systématique de ce genre et les méthodes d'investigation. – Annls Soc. ent. Fr. (N. S.) 18 (4): 483-506.
- 1986: Les diptères sarcophagidés orthoptères français – Essai biotaxonomique. – Université de Province, 301 pp.

Linnaeus, C., 1758: Systema naturae per regna tria naturae, secundum classes, ordines, genera, species, cum caracteribus, differentiis, synonymis, locis. Ed. 10. I, Holmiæ, 824 pp.

Loew, H., 1844: Beschreibung einiger neuen Gattungen der europäischen Dipterenfauna. – Stettin. ent. Ztg 5: 154-173.
- 1861: *Blaesoxipha grylloctona,* nov. gen., et spec. – Wien. ent. Mschr. 5: 384-387.

Lopes, H. S., 1959: A revision of Australian Sarcophagidae (Diptera). – Studia ent. 2: 33-67.
- 1982: On *Eumacronychia sternalis* Allen (Diptera, Sarcophagidae) with larvae living on eggs and hatchlings (sic!) of the east pacific green turtle. – Rev. bras. biol. 42 (2): 425-429.

Lundbeck, W., 1927: Diptera danica, genera and species of flies hitherto found in Denmark, part 7, Platypezidae and Tachinidae. – Copenhagen, 571 pp.

Lyneborg, L., 1960: Hanstedreservatets entomologi 9. Diptera: Brachycera & Cyclorrhapha –

Fluer. - Ent. Meddr 30: 201-262.

Macquart, J., 1835: Histoire Naturelle des Insectes. Diptères. 2. - Paris, 703 or 710 pp.
- 1843: Diptères exotiques nouveaux ou peu connus. - Mém. Soc. Sci. Agric. Lille 2 (3) (1842): 162-460, 36 pl. [reprinted with separate pagination 5-304].
- 1846: Diptères exotiques nouveaux ou peu connus. - Mém. Soc. Sci. Agric. Lille 1844: 133-164.
- 1850: Nouvelles observations sur les Diptères d'Europe de la tribu des Tachinaires (suite). - Annls Soc. ent. Fr. (2) 7 (1849): 357-418.
- 1854: Nouvelles observations sur les Diptères d'Europe de la tribu des Tachinaires. - Annls Soc. ent. Fr. (3) 2: 373-446.

Maneval, H., 1929: Observations sur *Hilarella stictica* Meig. (Dipt. Tachinidae) spoliatrice d'*Ammophila sabulosa* L. - Bull. Soc. ent. Fr. (1929): 26-28.

McAlpine, J. F., 1981: Morphology and terminology - adults. - Pp. 9-63 in McAlpine et al. (Eds): Manual of Nearctic Diptera 1. Biosystematics Research Institute, Ottawa.

Meade, R. H., 1876: Monograph upon the British species of *Sarcophaga* or flesh-flies. - Entomologist's mon. Mag. 12: 216-220, 260-268.

Meigen, J. W., 1803: Versuch einer neuen Gattungseintheilung der europäischen zweiflügeligen Insekten. - Mag. Insektenk. 2: 259-281.
- 1824: Systematische Beschreibung der bekannten europäischen zweiflügeligen Insekten. 4. - Hamm, xii + 428 pp.
- 1826: Systematische Beschreibung der bekannten europäischen zweiflügeligen Insekten. 5. - Hamm, xii + 412 pp.
- 1830: Systematische Beschreibung der bekannten europäischen zweiflügeligen Insekten. 6. - Hamm, xi + 401 pp.
- 1838: Systematische Beschreibung der bekannten europäischen zweiflügeligen Insekten. 7. - Hamm, xii + 434 pp.

Mihályi, F., 1965: Rearing flies from faeces and meat, infected under natural condition. - Acta zool. hung. 11 (1-2): 153-164.
- 1966: Flies visiting fruit and meat in an open-air market in Budapest. - Acta zool. hung. 12 (3-4): 331-337.
- 1979: Fémeslegyek - húslegyek, Calliphoridae - Sarcophagidae. - Fauna Hung. 135: 1-152.

Mik, J., 1890: Dipterologische Miscellen. 16. - Wien. ent. Ztg 9 (5): 153-158.

Miles, P. M., 1968: *Sarcophaga nigriventris* Meigen (Dipt., Calliphoridae) bred from *Helix aspersa* Müller (Mollusca, Helicidae). - Entomologist's mon. Mag. 104: 227.

Müller, A., 1922: Ueber der Bau des Penis der Tachinarier und seinen Wert für die Aufstellung des Stammbaumes und die Artdiagnose. - Arch. Naturgesch. 88A (2): 45-168.

Nielsen, E. T., 1932: Sur les habitudes hyménoptères aculéates solitaires I. (Bethylidae, Scoliidae, Cleptidae, Psammocharidae). - Ent. Meddr 18: 1-57.

Nielsen, J. C., 1914: Et angreb af sommerfuglelarver på et pilehegn. - Mindeskr. Japetus Steenstrups Føds. 1 (15): 1-9.

Nielsen, S. A., Nielsen, B. O. & Walhovd, H., 1978: Blowfly myiasis (Diptera: Calliphoridae, Sarcophagidae) in the hedgehog (*Erinaceus europaeus* L.). - Ent. Meddr 46: 92-94.

Nuorteva, M., 1946: Observations on the life of *Oxybelus uniglumis* L. (Hym., Sphegidae). - Ann. ent. fennici 11 (4): 213-217. (In Finnish with English summary).

Pandellé, L., 1896: Études sur les Muscides de France, II ᵉ partie. - Revue Ent. 15: 1-230.

Panzer, G. W. E., 1798: Fauna insectorum germanicae initia oder Deutschlands Insecten. Heft 54. - Nürnberg, 24 pp.

Pape, T., 1986: A revision of the Sarcophagidae (Diptera) described by J. C. Fabricius, C. F. Fallén, and J. W. Zetterstedt. - Ent. scand. 17: 301-312.
- in press: An annotated checklist of Finnish flesh-flies (Diptera: Sarcophagidae). - Notul. ent.

Park, S.-H., 1977: Studies on flies in Korea II. Taxonomical studies on sarcophagid flies (Diptera). – Bull. Tokyo med. dent. Univ. 24: 249-284.

Parmenter, L., 1950: *Blaesoxipha laticornis* (Mg.) (Dipt., Calliphoridae) as parasite of *Omocestus viridulus* (L.) (Orth., Acrididae). – Entomologist's mon. Mag. 86: 46.

Peckham, D. J., 1977: Reduction of miltogrammine cleptoparasitism by male *Oxybelus subulatus* (Hymenoptera, Sphecidae). – Ann. ent. Soc. Am. 70: 823-828.

Perris, D., 1852: Seconde excursion dans les Grandes-Landes. – Annls Soc. linn. Lyon 1850-1852: 145-216.

Pickens, L. G., 1981: The life history and predatory efficiency of *Ravinia lherminieri* (Diptera: Sarcophagidae) on the face fly (Diptera: Muscidae). – Can. Ent. 113: 523-526.

Pollock, J. N., 1972: Functional morphology of male genitalia in *Sarcophaga:* A comparative study. – Entomologist 105: 6-14.

Povolny, D. & Šustek, Z., 1983: Three dipterous representatives of the Carpathian fauna in the beech forests of central Moravia and the ecological preconditions of their discovery (Dipt., Sarcophagidae). – Sb. vys. Šk. zeměd. Praze 52 (1-2): 127-144.

Pyörnilä, M. & Pyörnilä, A., 1979: Role of parasitoids in termination of a mass occurrence of *Yponomeuta evonymella* (Lepidoptera, Yponomeutidae) in northern Finland. – Notul. ent. 59: 133-137.

Richards, O. W., 1960: A species of *Sarcophaga* (Dipt., Calliphoridae) new to Ireland. – Entomologist's mon. Mag. 96: 17.

– & Waloff, N., 1948: The hosts of four British Tachinidae. – Entomologist's mon. Mag. 84: 127.

Ringdahl, O., 1932: Eine neue *Brachycoma*-Art. – Notul. ent. 12: 21.

– 1937: Bidrag til kännedommen om de svenska tachinidernas utbredning. – Ent. Tidskr. 58: 31-38.

– 1945: Förteckning över de av Zetterstedt i Insecta Lapponica och Diptera Scandinaviæ beskrivna tachiniderna med synonymer jämte anteckningar över en del arter. – Opusc. ent. 10: 26-35.

– 1952: Catalogus Insectorum Sveciae XI. Diptera Cyclorrhapha: Muscaria Schizometopa. – Opusc. ent. 17: 129-186.

Ristich, S. S., 1956: The host relationships of a miltogrammid fly *Senotainia trilineata* (VDW). – Ohio J. Sci. 56: 271-274.

Roback, S. S., 1954: The evolution and taxonomy of the Sarcophaginae (Diptera, Sarcophagidae). – Illinois biol. Monogr. 23 (3-4): 1-181.

Robineau-Desvoidy, J. B., 1830: Essai sur les Myodaires. – Mém. prés. div. Sav. Acad. Sci. Inst. Fr. (2) 2: 1-813.

– 1863: Histoire naturelle des Diptères des environs de Paris. 2. – Paris, Leipzig, London, 920 pp.

Rognes, K., 1986a: The Sarcophagidae (Diptera) of Norway. – Fauna norv. ser. B, 33: 1-26.

– 1986b: The systematic position of the genus *Helicobosca* Bezzi with a discussion of the monophyly of the calyptrate families Calliphoridae, Rhinophoridae, Sarcophagidae and Tachinidae (Diptera). – Ent. scand. 17: 75-92.

Rohdendorf, B. B., 1930: 64h. Sarcophaginae. – Pp. 1-48 in Lindner, E. (Ed.): Die Fliegen der paläarktischen Region, Lieferung 39, Band 11, Stuttgart.

– 1934: Egyptian Larvivoridae collected by Prof. H. C. Efflatoun Bey (Diptera: Tachinidae). – Bull. Soc. ent. Égypte [1934] (1-2): 1-16.

– 1937: Fam. Sarcophagidae. – Fauna SSSR 19 (1): 1-501.

– 1955: Species of genus *Metopia* Mg. (Diptera, Sarcophagidae) from the fauna of USSR and adjacent countries. – Ént. Obozr. 34: 360-373. (In Russian).

– 1959: The species of Sarcophaginae flies in the faunistic synanthropic complexes of different landscape zones of the USSR. – Ent. Rev., Wash. 38: 708-714. (Translated from Ént. Obozr. 38: 790-797).

- 1963a: Über das System der Sarcophaginen der äthiopischen Fauna. - Stuttg. Beitr. Naturk. 124: 1-22.
- 1963b: Über wenig bekannte nordische Miltogrammatinen der Gattung *Oebalia* R.-D. (Diptera, Sarcophagidae). - Beitr. Ent. 13 (3-4): 445-454.
- 1965: Composition of the tribe Sarcophagini (Diptera, Sarcophagidae) in Eurasia. - Ent. Rev., Wash. 44: 397-406. (Translated from Ént. Obozr. 44: 676-695).
- 1967: The directions of the historical development of the sarcophagids (Diptera, Sarcophagidae). - Trudy paleont. Inst. 116: 1-92. (In Russian).

- 1969: New species of Sarcophaginae (Diptera, Sarcophagidae) from Asia. - Ent. Rev., Wash. 48: 600-604. (Translated from Ént. Obozr. 48: 943-950).
- 1970: Fam. Sarcophagidae - sarcophagidý. - Pp. 624-670 in Bei-Bienko, G. Ya. (Ed.): Opredelitel' nasekomýkh evropeisko chasti SSSR. 5. Akad. Nauk SSSR, Leningrad. (In Russian).
- & Verves, Yu.G., 1978: Sarcophaginae (Diptera, Sarcophagidae) from Mongolia. - Ann. hist.-nat. Mus. nat. hung. 70: 241-258.
- 1980: On the fauna of Sarcophagidae (Diptera) of the Mongolian People's Republic. 3. Miltogrammatinae. - Insects of Mongolia 7: 445-517. (In Russian).
Rondani, C., 1856: Dipterologiae Italicae Prodromus. 1. Genera Italicae ordinis Dipterorum ordinatim disposita et distincta et in familias et stirpes aggregate. - Parmae, 1-226 + 2 pp.
- 1859: Dipterologiae Italicae Prodromus. 3. Species italicae ordinis Dipterorum in genera characteribus definita, ordinatim collectae, methodo analitica distincta, et novis vel minus cognitis descriptis. Pars secunda. - Parmae, 1-243.
- 1860: Sarcophagae italicae observatae et distinctae. Commentarium XVIII pro Dipterologia italica. - Atti Soc. ital. Sci. nat. 3: 374-392.
- 1865: Diptera italica non vel minus cognita descripta vel annotata observationibus nonnulis additis. - Atti Soc. Ital. Sc. Nat., Milano 8: 193-231.
Roser, C. von, 1840: Erster Nachtrag zu dem in Jahre 1834 bekannt gemachten Verzeichnissen in Württemberg vorkommender zweiflügeliger Insekten. - CorrespBl. württ. landw. Ver. Stuttg. (N. S.) 17 (1): 49-64.
Rossi, P., 1790: Fauna Etrusca. Sistens insecta quae in provinciis Florentina et Pisana praesertim collegit. 2. - Liburni, 348 pp.
Saager, H., 1959: Die Dipterensammlung des Naturhistorischen Heimatmuseums der Hansestadt Lübeck. - Ber. naturhist. Mus. Lübeck 1: 21-62.
Saalas, U., 1943: *Parasarcophaga aratrix* Pand. (Dipt., Tachinidae) im Körper von *Prionus coriarius* entwickelt. - Ann. ent. fennici 9 (1): 23-28.
Sacca, G., 1945: Miasi da *Sarcophaga falculata* Pand. - Rc. Ist. sup. Sanità 8 (2): 301-302.
Sajo, K., 1898: Zur Lebensweise von *Sarcophila latifrons* Fall. und über Fliegen-Infektionen im allgemeinen. - Illte Z. Ent. 3: 149-151, 164-167.
Schiner, J. R., 1861: Vorläufiger Commentar zum dipterologischen Theile der "Fauna austriaca". 3. - Wien. ent. Mschr. 5 (5): 137-144.
- 1862: Fauna Austriaca. Die Fliegen ( Diptera). 1. - Wien, Lxxx + 674 pp.
Schmitz, H., 1917: Biologische Beziehungen zwischen Dipteren und Schnecken. - Biol. Zbl. 37: 24-43.
Schousboe, C., 1981: Forsøg over faktorers betydning for tre redeparasitters lokalisering af reder af humlebier (*Bombus* spp.) (Hymenoptera: Apidae). - Ent. Meddr 48 (3): 127-129.
Scopoli, I. A., 1763: Entomologia carniolica exhibens insecta carnioliae indigena et distributa in ordines, genera, species, varietates, methodo Linnaeana. - Vindobonae, 421 pp.
Séguy, E., 1921: Les Diptères qui vivent aux dépens des escargots. - Bull. Soc. ent. Fr. (1921): 238-239.
- 1932: Étude sur les diptères parasites ou predateurs des sauterelles. - Encycl. ent. sér. B. II. Dip-

tera 6: 11-40.
- 1941: Études sur les mouches parasites. 2, Calliphorides. Calliphorines (suite), sarcophagines et rhinophorines de l'Europe occidentale et meridionale. - Encycl. ent. (A) 21: 1-436.
- 1953: Diptères du Maroc. - Encycl. ent. sér. B. II. Diptera 11: 77-92.
- 1965: Le *Sarcophaga nigriventris* parasite de l'abeille domestique en Europe occidentale (Insect, diptère, calliphorid). - Bull. Mus. nat. Hist. natur. 37 (3): 407-411.
Senior-White, R., Aubertin, D. & Smart, J., 1940: Diptera. 6. Family Calliphoridae. - The Fauna of British India including the remainder of the Oriental region. London, xiii + 288 pp., 1 map.
Skaife, S. H., 1955: The black-mound termite of the Cape, *Amitermes atlanticus* Fuller. - Trans. R. Soc. S. afr. 34: 251-271.
Smith, K. G. V., 1957: Some miscellanous records of bred Diptera. - Entomologist's Rec. J. Var. 69: 214-216.
Smith, R. W., 1958: Parasites of nymphal and adult grasshoppers (Orthoptera: Acrididae) in western Canada. - Can. J. Zool. 36: 217-262.
Soper, R. S., Shewell, G. E. & Tyrell, D., 1976: *Colcondamyia auditrix* nov. sp. (Diptera: Sarcophagidae), a parasite which is attracted by the mating song of its host, *Okanagana rimosa* (Homoptera: Cicadidae). - Can. Ent. 108: 61-68.
Spofford, M. G. & Kurczewski, F. E., 1985: Courtship and mating behavior of *Phrosinella aurifacies* Downes (Diptera: Sarcophagidae: Miltogramminae). - Proc. ent. Soc. Wash. 87 (2): 273-282.
Sýchevskaya, V. I., 1970: Zonal distribution of coprophilous and schizophilous flies (Diptera) in middle Asia. - Ent. Rev., Wash. 49: 498-505. (Translated from Ént. Obozr. 49: 819-831).
Théodorides, J., 1954: Parasitisme de *Aeolopus strepens* (Latr.) (Orthoptera Acrididae) par *Pachyophthalmus signatus* Meigen (Diptera Calliphoridae) à Banyuls. - Vie Millieu 5: 457-458.
Thompson, P. H., 1978: Parasitism of adult *Tabanus subsimilis subsimilis* Bellardi (Diptera: Tabanidae) by a miltogrammine sarcophagid (Diptera: Sarcophagidae). - Proc. ent. Soc. Wash. 80 (1): 69-74.
Thompson, W. R., 1951: A catalogue of the parasites and predators of insect pests. Section 2, host parasite catalogue. Part 1, hosts of the Coleoptera and Diptera. - Commonwealth Agricultural Bureaux, Ottawa, 147 pp.
Thomson, C. G., 1869: 6. Diptera. Species nova descripsit. - K. svenska VetenskAkad. K. svenska fregatten Eugenies resa omkring jorden 2 (1): 443-614.
Tiensuu, L., 1939: Die Sarcophagiden (Dipt.) Finnlands. - Ann. ent. fennici 5 (4): 255-266.
Tolstova, Yu.S., 1962: On the insect fauna inhabiting the stems of raspberry (*Rubus idaeus* L.) in the suburban zone of Leningrad. - Ént. Obozr. 41: 285-293. (In Russian).
Townsend, C. H. T., 1892: Notes on North American Tachinidae s.str. with descriptions of new genera and species. Part 3. - Trans. Am. ent. Soc. 19: 88-132.
- 1937: Manual of Myiology 5. - Itaquaquecetuba, São Paulo, 232 pp.
- 1938: Manual of Myiology 6. - Itaquaquecetuba, São Paulo, 246 pp.
Venturi, F., 1953: Notulae dipterologicae 5. Revisione sistematica del genere *Metopia* Meigen (Diptera Sarcophagidae) in Italia. - Boll. Ist. Ent. Univ. Bologna 19: 147-170.
Verves, Yu.G., 1976a: The origin of inquilinism of miltogrammatin flies (Miltogrammatinae Sarcophagidae: Diptera). - Visn. kyiiv Univ., Ser. Biol. 18: 106-108. (In Russian with English summary).
- 1976b: The study of sarcophagids (Diptera, Sarcophagidae) - parasites of terrestrial gastropods. - Vestnik Zool. 1976 (3): 28. (In Russian).
- 1979: Description of *Paramacronychia hackmani* sp. n. (Diptera, Sarcophagidae, Paramacronychiinae). - Ann. ent. fennici 45 (1): 31-32.
- 1980: Karta 70 *Wohlfahrtia meigeni* (Schiner, 1862), Karta 71 *Sarcophaga carnaria* (Linnaeus,

1758), Karta 72 *Sarcophaga lehmanni* Müller, 1922. – In Gorodkov, K. B. (Ed.): Provisional atlas of insects of the European part of the USSR. Akad. Nauk SSSR, Leningrad. (In Russian).
- 1981a: Karta 122 *Ravinia striata* (Fabricius, 1794), Karta 123 *Sarcophaga schulzi* Müller, 1922, Karta 124 *Sarcophaga subvicina* Rohdendorf, 1937. – In Gorodkov, K. B. (Ed.): Provisional atlas of insects of the European part of the USSR. Akad. Nauk SSSR, Leningrad. (In Russian).
- 1981b: Sarcophagidae (Diptera) from Korea. – Folia ent. hung. 34 (1): 197-201.
- 1982a: New data on taxonomy of Sarcophagidae (Diptera). – Ént. Obozr. 61 (1): 188-189. (In Russian).
- 1982b: 64h. Sarcophaginae. – Pp. 235-296 in Lindner, E. (Ed.): Die Fliegen der palaearktischen Region. Lieferung 327, Band 11, Stuttgart.
- 1983: The American species of the genus *Macronychia* Rondani (Macronychiinae Sarcophagidae, Diptera). – Rev. brasil. Biol. 43: 345-354.
- 1985: 64h. Sarcophaginae. – Pp. 297-440 in Lindner, E. (Ed.): Die Fliegen der palaearktischen Region. Lieferung 330, Band 11, Stuttgart.
- 1986: Family Sarcophagidae. – Pp. 58-193 in Á. Soós (Ed.): Catalogue of Palaearctic Diptera, Vol. 12, Calliphoridae – Sarcophagidae. Budapest.
- & Kuz'movich, L. G., 1979: Sarcophagines (Diptera, Sarcophaginae) – parasites of terrestrial gastropods in the Ternopol region. – Vestnik Zool. 1979 (4): 16-21. (In Russian with English summary).
Viktorov-Nabokov, O. B. & Verves, Yu. G., 1975: K izucheniyu mukh (Diptera: Calliphoridae, Sarcophagidae), parazitiruyushchikh v dozhdevýkh chervyak (Oligochaeta, Lumbricidae). – Problemý pochvennoi Zoologii (Vil'nyus) 1975: 97-98.
Villeneuve, J., 1899: Description de Diptères nouveaux. – Bull. Soc. ent. Fr. 1899: 26-28.
- 1911: Dipterologische Sammelreise nach Korsika. (Dipt.) Tachinidae. – Dt. ent. Z. 1911: 117-130.
- 1912a: Sarcophagides nouveaux. – Annls hist.-nat. Mus. natn. hung. 10: 610-616.
- 1912b: Diptères nouveaux du nord Africain. – Bull. Mus. natn. Hist. nat. Paris 1912: 415-417, 505-511.
- 1922: Myodaires supérieurs paléarctiques nouveaux. – Annls Sci. nat. Zool. (10) 5: 337-342.
Walker, F., 1859-1860: Catalogue of the dipterous insects collected at Makessar in Celebes, by Mr. A. R. Wallace, with descriptions of new species. – J. Proc. Linn. Soc. Lond., Zool. 4: 90-96, 97-144 (1859), 145-172 (1860).
Walton, W. R., 1915: Report on some parasitic and predaceous Diptera from northeastern New Mexico. – Proc. U.S. natn. Mus. 48: 171-186, 2 pl.
Weis, F., 1960: Usædvanligt fund i en seng. – Ent. Meddr 29: 376-377.
Wood, D. H., 1933: Notes on some dipterous parasites of *Schistocerca* and *Locusta* in the Sudan. – Bull. ent. Res. 24: 521-530.
Zetterstedt, J. W., 1838: Sectio tertia. Diptera. Dipterologis scandinaviae, amicis et popularibus carissimis. – Pp. 477-868 in: Insecta lapponica "1840", Leipzig.
- 1844: Diptera scandinaviæ disposita et descripta. 3. – Lundæ, 895-1280.
- 1845: Diptera scandinaviæ disposita et descripta. 4. – Lundæ, 1281-1738.
- 1855: Diptera scandinaviæ disposita et descripta. 12. – Lundæ, xx + 4547-4942.
- 1859: Diptera scandinaviæ disposita et descripta. 13. – Lundæ, xvi + 4943-6190.
Zhang, M., 1982: A study of the larvae of some common sarcophagid flies from China. – Entomotaxonomia 4 (1-2): 93-106. (In Chinese with English summary).
Zumpt, F., 1965: Myiasis in man and animals in the Old World. – London, xv + 267 pp.
- 1972: Calliphoridae (Diptera Cyclorrhapha). Part 4: Sarcophaginae. – Explor. Parc natn. Virunga. Miss. G. F. de Witte 101: 1-264.

# Index

Synonyms are given in italics. The number in bold refers to the main treatment of the taxon.

201